Lecture Notes in Mathematics

Edited by A. Dold and B. Eckmann

1086

Sensitivity of Functionals with Applications to Engineering Sciences

Proceedings of a Special Session of the American Mathematical Society Spring Meeting held in New York City, May 1983

Edited by V. Komkov

Springer-Verlag
Berlin Heidelberg New York Tokyo 1984

Editor

Vadim Komkov
Department of Mathematics, Winthrop College
Rock Hill, SC 29733, USA

AMS Subject Classification (1980): 49H, 73K, 73L

ISBN 3-540-13871-4 Springer-Verlag Berlin Heidelberg New York Tokyo
ISBN 0-387-13871-4 Springer-Verlag New York Heidelberg Berlin Tokyo

Printing and binding: Beltz Offsetdruck, Hemsbach / Bergstr.
2146/3140-543210

Introduction

This volume contains an enlarged version of some of the talks presented at the American Mathematical Society Spring Meeting in New York City, May of 1983 in a special session on the Sensitivity of Functionals with Applications to Engineering Sciences. All talks were given by invitation. The enlarged versions of the talks given by professors M. Vogelius and R.V. Kohn, and by A. Fiacco and J. Kyparisis were published elsewhere before the publication of this volume. Only an enlarged abstract exists of Professor W.M. Wonham's presentation. Versions of all other papers presented at that meeting are given in this volume. All of these papers discuss applications of sensitivity theory to design of engineering systems or processes. Applications of sensitivity theory to chemical kinetics are given in the contribution of Professor H. Rabitz.

Applications to structural or mechnaical engineering were offered in the papers of V. Komkov; of J.W. Hou, E.J. Haug and R.L. Benedict, A. Diaz, N. Kikuchi and J.E. Taylor, V. Komkov and C. Irwin and P. Pedersen. All of these papers stress the functional analytic approach to optimization and to sensitivity theory for structural and mechanical systems.

These contributions attempt to clarify some latest developments in sensitivity theory for engineering systems and also present new theoretical solutions to several important problems. For example, the paper of H. Rabitz presents a new insight into the numerical treatment of chemical kinetics and of quantum scattering theory based on the sensitivity density concept. The paper of P. Pedersen deals with the troublesome aspects of nonselfadjoint problems that arise in structural dynamics, such as the dynamic behavior of the Beck column or Hauger column with both external and internal damping. Influence of these theoretical findings on existing numerical techniques is clearly outlined. Grid modification problems for numerical solutions of structural optimization are the main topic of the paper of A.R. Diaz, N. Kikuchi and J. Taylor. The modification of the grid and related sensitivity with respect to numerical errors is the main topic of that paper. It develops a new direction in numerical implementations of optimizing procedures, but more importantly it provides rigorous mathematical justification for such procedures. This aspect has been lacking in almost all previous papers on this subject.

An important mathematical development in structural and mechanical optimization is discussed in the J.W. Hou, E.J. Haug and R.L. Benedict paper on the shape optimization. While the specific problem concerns optimization of multiply-connected elastic bars subjected to pure torsion, the basic concepts introduced in that paper are quite general. The authors use effectively the idea of material derivative in the so called "speed method" that was originated by French mathematicians to formulate the sensitivity of a cost functional with respect to admissible design changes.

This topic is also discussed in the paper of V. Komkov where he suggests among other topics a group-theoretic formalism borrowed from quantum mechanics to encompass the purely functional-analytic approach now prevailing in the optimization of design and in the sensitivity analysis techniques now in use for designing of structural systems.

A different approach to identification of nondifferentiability or bifurcation is developed in the paper of V. Komkov and C. Irwin, based on embedding the optimization of design problems in a higher dimensional design space. An algorithm suggested in that paper avoids the usual pitfalls of gradient-type approach to optimization of design, where the difficulties are caused by the bifurcation phenomena.

While the contents of all papers can be safely described as applied mathematics, one of the primary purposes of the theories developed here is the advancement of technology, particularly of the C.A.D. (computer assisted design) technology.

The recent increase in the research activities concerning optimal design and the sensitivity of functionals with respect to design parameters can be directly linked to the advances in computer technology and to the recent growth in the C.A.D., and C.A.M. applications to the industrial and scientific processes.

TABLE OF CONTENTS

Vadim Komkov
Sensitivity Analysis in Engineering Applications.
More Specifically, in Civil and Mechanical Engineering Applications 1

Jean W. Hou, Edward J. Haug and Robert L. Benedict
Shape Optimization of Elastic Bars in Torsion 31

Alejandro Diaz, Noboru Kikuchi and J.E. Taylor
Optimal Design Formulations for Finite Element Grid Adaption 56

Herschel Rabitz
Sensitivity Methods for Mathematical Modelling 77

Vadim Komkov and Carlton Irwin
Uniqueness for Gradent Methods in Engineering Optimization 93

Pauli Pedersen
Sensitivity Analysis for Non-selfadjoint Problems 119

SENSITIVITY ANALYSIS IN SOME ENGINEERING APPLICATIONS

More Specifically in Civil and Mechanical Engineering Applications

Vadim Komkov
Department of Mathematics
Winthrop College
Rock Hill, S.C. 29733

ABSTRACT

In this very concise presentation we outline some historical developments of optimization theories as applied to continuum mechanics and to mechanical and civil engineering designs and trace the development of various modern sensitivity techniques during the period of preceding twenty years. We outline some of the difficulties and the progress made in overcoming them. We also stress some of the recently developed theoretical methods such as the "speed" method, and group theoretic techniques,indicating their importance to the computer-aided technology. Finally, we briefly outline some possible future directions.

We briefly discuss possible future connections between the design optimization theory and control theory.

ACKNOWLEDGEMENT

This research was supported by the NSF Grant # CMS 80-05677. The author also acknowledges the use of the facilities at the University of Iowa, College of engineering, Iowa City, Iowa 52240.

1. INTRODUCTION AND A BRIEF HISTORY OF THE PROBLEM.

In recent years massive volume of new results on optimization of systems with distributed parameters appeared in both mathematical and engineering journals. The engineering insight into some fairly complex problems helped greatly to point out very serious difficulties which sometimes were completely overlooked in "purely mathematical" articles. Difficulties which prevent an engineer from designing a system are sometimes overlooked, if the mathematical formulation is such that the "ugly cases" are excluded. This can be accomplished by assumming conditions convenient for the proofs of theorems. Many structural and mechanical problems studied in the "mainstream" of engineering design practice are nonlinear and display disturbing behavior that we called "ugly". For a while some practical engineers were loosing faith in the predicted "optimal" designs derived by the generally accepted numerical methods, such as the gradient, or gradient-projection algorithms, and insisted on validity of heuristically derived results that were based only on physical intuition.

The progress of theoretical work which could be immediately applied to structural and mechanical problems was hardly at all related to the corresponding mathematical developments in the theory of optimization. The cases, where loss of stability and bifurcation phenomena could be expected, were meeting with unexpected difficulties. The simplest one-dimensional problem of structural stability, namely the optimization of a slender column has a long history. In 1770 Lagrange wrote a paper on optimization of the shape of columns against buckling. The conclusions were generally incorrect. (See [1] for a brief historical outline and [2] for the original document). For a long time there was little progress in the engineering design theory and very little progress in design optimization or sensitivity, despite the great advances attained in calculus of variations and in related areas of mathematical analysis. New directions became available in the 1950s following the advances in functional analysis, and rapid development in control theory. A number of authors realized that control-theoretic results may be directly applied to some problems of engineering optimization. For example, in an obscure publication, E.J. Haug [6] reviews his own approach of converting some problems of optimal design to the equivalent problems of control theory. Here we offer a similar slightly different version. Consider the optimization of the volume of a column:

$$J = \int_0^L A(\underset{\sim}{u}(s))\, ds \rightarrow \min,$$

where $\underset{\sim}{u}(s)$ is the vector of design parameters, $A(s)$ is the magnitude of the cross-sectional area, ds denotes the element of length measured along the shape of the column. The state equation is given by:

$$EI(\underset{\sim}{u}(s))y''(s) + Py = 0,$$

where E is Young's modulus, I the moment of inertia of the cross-section about the neutral axis of bending, and P is the applied load. The boundary conditions at s = 0 are

$$y(0) = 0,$$

$$y'(0) = 0.$$

A constraint imposed on the design vector $\underset{\sim}{u}$ is

$$P/A(\underset{\sim}{u}) < \sigma_{max}$$

where σ_{max} is the given maximal stress level.

The functions $A(\underset{\sim}{u}(s))$, $I(\underset{\sim}{u}(s))$ are positive and piecewise continuous.

Let us change our notation $x_1 = y$, $x_2 = \frac{dy}{ds}$. Then the state equation can be rewritten as a system

$$1.1)\quad \begin{aligned} \dot{x}_1 &= x_2 = f_1(s), \\ \dot{x}_2 &= -Px_1/(EI(\underset{\sim}{u}(s)) = f_2 . \end{aligned}$$

After introducing a dual variable λ and a Hamiltonian,

$$H = \begin{cases} H_1 = -\lambda_0 A(\underset{\sim}{u}(s)) + \lambda_1 x_2 - \lambda_2\, Px_1/(EI(\underset{\sim}{u}(s))) - \mu(P/A(\underset{\sim}{u}) - \sigma_{max}) \\ H_2 = \lambda_1 x_1(o) + \lambda_2 x_2(o), \end{cases}$$

one can apply directly Pontryagin's maximality principle to formulate the necessary conditions for optimality of the design $\underset{\sim}{u}(s)$.

While a number of publications appeared following this basic approach (see for example Armand [8] or de Silva [7], somehow the basic difficulty of design theory has not been resolved by such control-theoretic techniques. To illustrate the basic difference, let us offer a physical example. Let us consider a single loop, distributed parameter L-C network. The differential equation modeling its behavior is given by the "telegraph" system of equations

$$1.2)\quad \begin{aligned} L(x)\frac{di}{dt} + \frac{dv}{dx} &= f_1(x,t) \\ C(x)\frac{dv}{dt} + \frac{di}{dx} &= f_2(x,t) \end{aligned}$$

Here $f_1(x,t)$, $f_2(x,t)$ denote, respectively, the voltage and current sources that are applied externally to the network. In control theory one usually accepts the general form of the differential equations modeling the system, that is the basic <u>design</u> of the system and adjusts the (external) input into the system to optimize some a priori given performance criterion.

Rewriting (1.2) as a vector equation we have

$$E(x)\,\frac{\partial \underset{\sim}{w}}{\partial t} + A\,\frac{\partial \underset{\sim}{w}}{\partial x} = \underset{\sim}{u}\,(x,t),$$

with

$$\underset{\sim}{w} = \begin{bmatrix} i \\ v \end{bmatrix},\; E(x) = \begin{bmatrix} L(x) & 0 \\ 0 & C\,(x) \end{bmatrix},\; A = \begin{bmatrix} 0 & 1 \\ 1 & 0 \end{bmatrix},\; \underset{\sim}{u}(x,t) = \begin{bmatrix} f_1\;(x,t) \\ f_2\;(x,t) \end{bmatrix} .$$

Control problem consists of selecting a vector $\underset{\sim}{u}(x,t)$ in a space U of admissible controls such that the state of the system $\underset{\sim}{w}$ and the control $\underset{\sim}{u}$ optimize some performance criterion. The system may be an open loop system as defined in this example or it may be a closed loop feedback system. However, the basic equation defining the state of an open loop system is regarded as given. On the other hand in a design optimization problem one considers possible changes in the model of the system, i.e., replacement of the operator, $E(x)\frac{\partial}{\partial t} + A\frac{\partial}{\partial x}$ acting a vector $\underset{\sim}{w} = \begin{bmatrix} i \\ v \end{bmatrix}$ in a Sobolev space $H^1(\Omega)$ by a different operator possibly mapping different Sobolev spaces into each other, as well as possible changes in a feedback design. One could, for example, consider changes in $E(x)$, i.e. changes made by adjusting only the distributed inductance and capacitance. Or, one could introduce a resistance, altering completely the form of the differential operator and distributing that resistance along the wire. Design changes would incorporate the changes in this resistance.

Perhaps, a simple problem of design improvement consists of changes in the coefficients of the differential, or integrodifferential operator defining the state of an open loop system. In some cases involving feedback systems it is hard to separate the control and the design considerations and to compartamentalize such procedures. Nevertheless, the reasons why control-theoretic techniques have not been used successfully to solve or at least to approach numerically solutions to design problems are understood. The design optimization represents a higher level of difficulty. Combined optimization of design and of control which is of extreme importance in space engineering and "star wars" weapon technology has been treated in purely heuristic terms in technical reports and only a few papers appeared which made a serious attempt at theoretical principles which must govern such general class of problems (see [9], [10]).

I can predict that this class of problems will be extensively investigated in the near future basing such predictions on the remarks of experts in space technology, who stress the urgent need of such "unified" approach to designs of space vehicles and "superweapons". However, this theoretical development is still in the future.

The development of control theory in the 1950s and 1960s gave some impetus to design considerations, but the serious revival of interest in engineering design theory can be credited to a series of articles of J.B. Keller analyzing the one-dimensional case involving the loss of stability. In an elegant paper [3] published in 1960 Keller analyzed the distribution of the cross-sectional area of a column. He used both the classical theory of ordinary differential equations and the modern application of directional derivatives to derive some necessary conditions for the design optimality. In [4] Tadjbakhsh and Keller continued this development extending Keller's results to different types of boundary conditions. Subsequent papers of Keller and Niordson [11], Farshad and Tadjbakhsh [12], and others (see for example [13]) pursued the general approach of Keller. However, numerical results based on these techniques lead to the formation of physically unexplainable singularities. Such singular solutions persisted in the numerical optimization for structural and mechanical problems. Results based on the theory given in [3], [4],

[5] exhibited singularities. A simple check of some optimal designs revealed that some error must be present. The error made in the Tadjbakhsh and Keller paper [4] was a subtle one. The authors assumed the existence and continuity of the first derivative. These difficulties lead to a speculation that a constraint should be imposed on the minimum size of the cross-sectional area of the buckled column to avoid formation of singularities. A typical optimal design containing singularities, copied from N. Olhoff's thesis, is shown below in Figure 1.

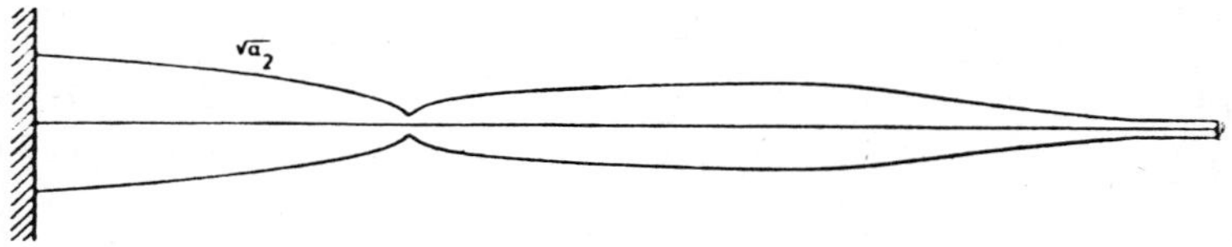

Optimal λ_2 cantilevers without nonstructural mass. The solution above corresponds to $\bar{\alpha} = \bar{A}L/V = .05$, and has $\lambda_2 = \omega_2^2 \rho L^5/cEV = 1.74 \times 10^3$, $\omega_2/\omega_2^* = 1.89$, $\Delta/\Delta^* = 2.11$. The solution below: $\bar{\alpha} = .5$; $\lambda_2 = 8.96 \times 10^2$, $\omega_2/\omega_2^* = 1.36$, $\Delta/\Delta^* = 1.40$.

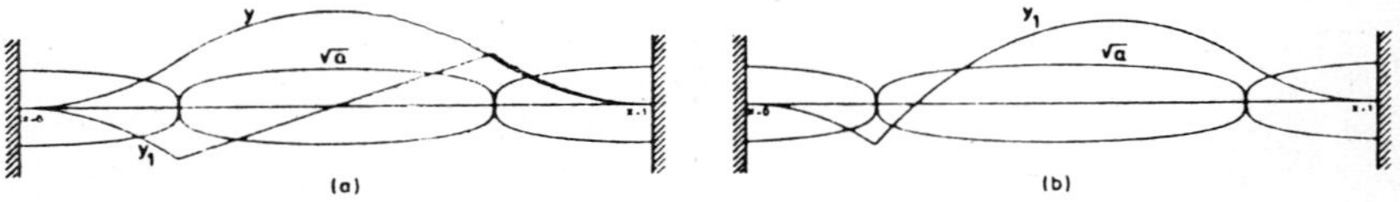

Geometrically unconstrained doubly clamped column designs obtained by single mode formulations. (a) Incorrect design $\pm\sqrt{\alpha}$ and mode y determined in [27]. The true fundamental mode of the design is indicated by y_1. The points of zero thickness are located at distances L/4 from column ends. (b) Correct design $\pm\sqrt{\alpha}$ and corresponding fundamental mode y_1 of a single mode formulation. Zero thickness points are optimally located at distances 0.208 L from column ends.

Figure 1

Olhoff's solutions for constrained and unconstrained vibrating cantilever and clamped-clamped column, respectively.

A more realistic optimal shape obtained by Olhoff [5] with constraints imposed on minimal cross-section is shown in Figure 2.

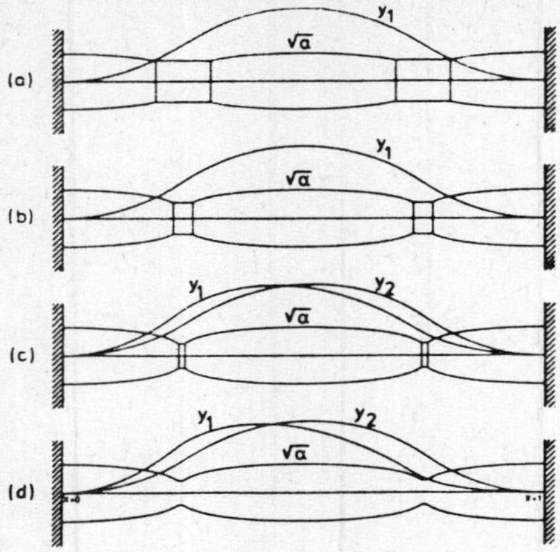

Optimal column designs $\pm\sqrt{\alpha}$ and associated fundamental modes subject to different values of geometric minimum constraint $\overline{\alpha}$: (a) $\overline{\alpha} = 0.7$, $\lambda = 48.690$ is simple. (b) $\overline{\alpha} = 0.4$, $\lambda = 51.775$ is simple. (c) $\overline{\alpha} = 0.25$, $\lambda = 52.349$ is double, and the geometric minimum constraint is still active. (d) The $\overline{\alpha}$ independent, optimal design for $0 \leq \overline{\alpha} < 0.226$ (with inactive minimum constraint). $\lambda = 52.3563$ is double.

Figure 2

Various designs including singularities were analyzed by Olhoff and Taylor [14], Masur [12], and by Mróz and Rozvany [15] who realized that such designs could be regarded as rather poor approximations but could be improved by locating structural supports at near-singular points.

While singularities persisted, one could easily understand the wise advice given by W. Prager that it is best not to get carried away when one is optimizing. That is, one should optimize for awhile then quit. This was a purely heuristic statement which was well supported by numerical findings. A number of papers appeared in the engineering literature which remedied this unfortunate situation by assigning lower (or upper) bounds on design variables, thus preventing the onset of singularities in the "optimal" solution. This was shown to be unnecessary in an interesting but basically formal computation carried out by Olhoff and Rasmussen (see [5], [30]), who demonstrated that such inequality constraints become slack if bimodal optimization analysis is carried our correctly.

Similar problems have arisen in the studies of dymanic systems. In his talk [46] given at A.M.S. meeting in New York, NY, (1983), P. Pedersen offered an interesting analysis of a very difficult phenomenon which could be classified as a form of flutter which arose naturally in the sensitivity analysis of dynamic behavior of structures. This behavior consists of the vibrating structure changing its mode from the fundamental mode to the second eigenmode and back again when a specific design is adopted.

The results quoted here were obtained by studying sensitivity, rather than concentrating on optimality of design with respect to some (fixed) eigenmode. At the same time this analysis reveals that inherent difficulties are present in following a purely formal approach to the optimization of design.

A study of structural optimization under general transient response to dynamic loads was conducted in a series of papers of E.J. Haug and his associates at the University of Iowa [31], [32], [33], [34], [35], [36], where some of these difficulties were discussed. In [36], these authors demonstrated that some previously obtained results on optimization were at best suboptimal and, in general, were not optimal designs. Using sensitivity analysis more carefully these authors applied the general theory with carefully selected constraints to derive a most interesting designs of vehicle suspension systems, power distribution towers and a completely unintuitive design of a plastic sabot for a large caliber weapon system.

Developments in structural optimization analysis in the Soviet Union were strongly influenced by modern functional analysis. For example papers of V.G. Litvinov [25], Kryśko [26], Seyranian and Gura [27], [28], [29] contain sophisticated Sobolev space arguments proving the existence of weakly converging optimizing sequences in various problems of optimal design. In particular the papers [27], [28], [29] contain important theoretical results which reinforce the theory of quasi-differentiability applied to the structural buckling and vibration problems, as pioneered by Haug, Rousselet, Choi. Other results establishing existence and stability of optimal design and of weak convergence based on sensitivity sequences were given by Velte and Villagio in [36].

2. MULTIPLE EIGENVALUES AND BIFURCATION.

Full implications of merging of eigenvalues are still not understood. The following two-dimensional (i.e., with two degrees of freedom) mechanical system was offered as an example of problems which arise in the infinite dimensional problems of design and sensitivity optimization. (See [24]). Moreover in this simple setting one can comprehend the physical reasons for certain phenomena which are quite mistifying in the infinite dimensional case. Consider the vibration of a rigid bar with uniformly distributed mass and additional point mass located at a distance "a" from one end. The bar is supported by two linear springs.

This example is discussed in detail in the paper of V. Komkov and C. Irwin in this issue. It suffices to point out that the following listed phenomena-that occur simultaneously at the optimal distance "$\hat{a}$"-determine the onset of bifurcation and form a barrier to an orderly gradient optimization approach.

At the point of optimal design the following phenomena occur:

1) The design is "truly" optimal.

2) The functional Λ is not differentiable.

3) The Euler-Lagrange equations of motion become decoupled.

4) The dimension of the null space of the state operator changes.

5) The boundary value problem for the state operator becomes ill-imposed in the sense of Hadamard.

We could add

6) A suitable embedding in a higher dimensional space reveals that the value of the design variable $\hat{a} = \frac{\ell}{2}$ represents a unique point causing certain symmetry arguments to break down.

This last remark need to be explained in more detail. It is the basis of a substantial part of the paper [24] by the author and C. Irwin.

The idea of embedding the class of design problems in a larger space is attractive from many points of view. While in [32] additional parameter (length of column) was chosen successfully for prediction of bifurcation phenomena and for deriving an estimate of eigenvalues in the one-dimensional case, in the more general case such choice is not obvious. Group theoretic arguments suggest themselves as a possible way of selecting additional parameters. For a general discussion of such approach see [22] or the classical work of Ovsiannikov [38]. For application to the problems in optimal design of columns see [37]. This approach, which tries to parallel some well-known applications of Lie groups to theoretical physics is only in the initial stage of its development but it is very promising in the opinion of the author.

3. THE TREATMENT OF CONSTRAINTS AND COST FUNCTIONALS.

In most of the theoretical papers on design optimization published during the 1967 - 1977 period constraints were handled in the traditional manner by the use of Lagrangian multipliers. (For example, see [5], [6], [14],[15].) In some cases, where lack of smoothness, or continuity caused difficulties, clever ad hoc techniques were devised to incorporate the constraints into the optimization algorithms. William Prager, John Taylor and other derived ingeneous techniques of bypassing the Lagrangian multiplier method ([40], [41], [9]). For example,the Betti-Castigliano formula was used in design problems with constraints on deflection at a point or on total complience for the Euler-Bernoulli, or Euler-Timoshenko beams. The difficulty one encounters with this rather clever approach is apparent when one tries to generalize it to higher dimensional cases (plates or shells) or to statically indeterminate structures.

Other authors adopted a basically heuristic approach to these problems. While papers such as [5], [6] were physically reasonable, they lacked mathematical rigor. A rigorous discussion which parallels some ideas of control theory, but in a novel setting, was advanced by Haug and Komkov [42], and developed in greater detail in [34], [35], [13] by Haug and associates at the University of Iowa.

This type of problem discussed in these papers is best explained in a Hilbert space setting. The state of the system is given by a system of linear differential equations

3.1) $L(\underset{\sim}{u})\underset{\sim}{z} \;=\; Q(\underset{\sim}{x},\ \underset{\sim}{z}(\underset{\sim}{x}),\ \underset{\sim}{u}),$ where $\underset{\sim}{x} \in \Omega \subset R^n$

is a local coordinate system. An operator L maps a Hilbert space H_1 into a Hilbert space H_3, $z(\underset{\sim}{x})$ is the state vector in H_1, $\underset{\sim}{u}$ is design vector which belongs to space of admissible designs U whose topology is determined by the physics of the problem. (In fact, careful definition of U is an important part of the mathematical modelling.

Boundary conditions are determined by a set of equations

3.1a) $\hat{B}\underset{\sim}{z} = q(\underset{\sim}{x})\ ,\quad \underset{\sim}{x} \in \partial\,\Omega\ .$

In many problems of structural mechanics the operator

$$A = \begin{bmatrix} L & 0 \\ 0 & B \end{bmatrix}:\ H_1 \times H_2 \rightarrow H_3 \times H_4$$

is positive and bounded below and A^{-1} is a completely continous operator. In such cases a fairly straightforward approach consisting of iterative application of a gradient type approach is justified in design improvements involving minimization of a functional

$$\psi_0:\ [\mathbb{R}^m \times H_1 \times H_2 \times U] \rightarrow \mathbb{R}.$$

Generally, state and performance constraints are assigned, such as

3.1b) $\psi_\alpha \;=\; h_\alpha(\underset{\sim}{u},\ \underset{\sim}{x},\ 5(\underset{\sim}{x},\underset{\sim}{u})) + \int_\Omega g_\alpha(x,\underset{\sim}{u},z(x))dx + \int_{\partial\Omega} f_\alpha(\underset{\sim}{z},\underset{\sim}{x})\ dx \leq 0,$

$\alpha = 1, 2, \ldots r.$

The basic "trick" which was introduced in [42], [35] consists in completely by-passing the computation of the sensitivity of the cost functional ψ_0 with respect to the state variable $\underset{\sim}{z}$, while incorporating the constraint conditions (3.1b) into the sensitivity formula.

This is best explained on the linear example. The state equations

$$L\underset{\sim}{z} \;=\; Q(\underset{\sim}{x})\ , \qquad \underset{\sim}{x} \in \Omega \quad \text{, and}$$

$$\hat{B}\underset{\sim}{z} \;=\; q(\underset{\sim}{x})\ , \qquad \underset{\sim}{x} \in \partial\Omega \quad ,$$

are replaced by variational arguments

3.2) $a(\underset{\sim}{z},\lambda,\underset{\sim}{u}) \;=\; \langle L\underset{\sim}{z},\underset{\sim}{\lambda}\rangle_\Omega + \langle \hat{B}\underset{\sim}{z},\hat{\underset{\sim}{\lambda}}\rangle_{\partial\Omega} - \langle Q(\underset{\sim}{x}),\underset{\sim}{\lambda}\rangle_\Omega - \langle q,\lambda\rangle_{\partial\Omega} + \sum_{i=0}^{r} \mu_i\ \psi_i(\underset{\sim}{z}) \;=\; 0.$

In appropriate Sobolëv spaces this equation remains invariant under admissible design changes. Let us suppose, for the sake of simplicity, that the loads applied $(Q(\underset{\sim}{x}))$ are independent of the design and independent of the state $\underset{\sim}{z}$. Hence, for the sake of simplicity, let us ignore these conditions in (3.2).

If $\underset{\sim}{z}$ is Fréchet differentiable function in the Sobolëv space H assigned to this problem and L and B are operators which depend smoothly on the design $\underset{\sim}{u}$, one can repeat a formal treatment of the problem encountered in many engineering papers of perturbing $a(\underset{\sim}{z},\underset{\sim}{\lambda})$ by varying the design vector $\underset{\sim}{u}$. One could introduce some rigor into this approach by checking Kato's conditions for existence of abstract derivatives of the operators L and B that map appropriate Hilbert spaces into each other. The details of this theoretical justification may be found in the Haug-Komkov paper [42]. Further, one could check the differentiability of the constraints and justify the use of the Lagrangian multipliers. Instead, let us pursue the main thrust of the Haug-Arora-Komkov "trick" which incredibly simplifies the computational work and the corresponding programming task. It should be observed that even if we were justified in formally differentiating the functional $a(\underset{\sim}{z},\underset{\sim}{\lambda})$ with respect to the design $\underset{\sim}{u}$, thus obtaining a formula for the sensitivity of the cost functional and of the constraint functionals, such formula would contain terms which feature abstract derivatives $\frac{\partial \underset{\sim}{z}}{\partial \underset{\sim}{u}}$ (or $\frac{\partial \underset{\sim}{z}}{\partial \underset{\sim}{u}} \cdot \delta u$ in the more common engineering "δ" - symbolism). That is, to find the sensitivity of the cost functional one must know the sensitivity of the state of the system. This would usually imply a numerical scheme in which the state of the system is repeatedly computed at each step and the sensitivity of the state with respect to the design is computed for each change in design.

Moreover, such elaborate configurations must include the check on admissibility, that is, the state variables must obey all stated conditions and constraints while the design (before and after the change) must belong to an a priori stated set of admissible designs. In realistic design computations, where the design vector may include hundreds of design functions and many lumped parameters, such computations are lengthy, difficult to program, and overload the memory of even modern, medium size computers. The "adjoint variable" technique relies on the fact that the dual variable $\underset{\sim}{\lambda}$ (in the variational formulation (3.2)) does not have to satisfy any a priori ordered boundary or initial conditions. In fact, we postulate that (3.2) may be satisfied by certain class of functions $\underset{\sim}{\lambda}(\underset{\sim}{x})$ in a Sobolëv space determined by the physics of our problem.

Suppose that $\underset{\sim}{\lambda}$ is a weak solution to the differential equation

3.3) $$L^*\underset{\sim}{\lambda} = \frac{\partial \Sigma \mu_i \psi_i(\underset{\sim}{z})}{\partial \underset{\sim}{z}}$$ (where $\frac{\partial \underset{\sim}{\psi}}{\partial \underset{\sim}{z}}$ has to be interpretated as a tensor product),

with the simplest available boundary and finite conditions. For example, $\lambda \equiv 0$ on $\partial\Omega$ and $\underset{\sim}{\lambda}(t = T) = 0$ in the dynamic case in which the time interval is $[0,T]$. Specifically, let $\underset{\sim}{\psi} \equiv \psi_0$ (scalar case).

We can use the trivial identity

$$\frac{d\psi_o(\underset{\sim}{z},\underset{\sim}{u})}{d\underset{\sim}{u}} = \left< \frac{\partial\psi_o}{\partial\underset{\sim}{z}} , \frac{\partial z}{\partial\underset{\sim}{u}} \right> + \frac{\partial\psi}{\partial\underset{\sim}{u}} \quad (\in U^*) \quad ,$$

where U* is the topological dual of space containing admissible designs U. Hence, we could write

$$3.3^a) \quad \left< L^*\underset{\sim}{\lambda} \right> \frac{\partial\underset{\sim}{z}}{\partial\underset{\sim}{u}} = \left< \frac{\partial\psi_o}{\partial\underset{\sim}{z}} , \frac{\partial\underset{\sim}{z}}{\partial\underset{\sim}{u}} \right> = \frac{d\psi_o}{d\underset{\sim}{u}} - \frac{\partial\psi_o}{\partial\underset{\sim}{u}} \quad \in U^* .$$

On the other hand,

$< L^*\underset{\sim}{\lambda}, \underset{\sim}{z} >$ is design independent in our simple case, since

$< L^*\underset{\sim}{\lambda}, \underset{\sim}{z} > \; = \; <\underset{\sim}{\lambda}, L\underset{\sim}{z} > \; = \; <Q(\underset{\sim}{x})\underset{\sim}{\lambda} >$

Hence,

$$\frac{d}{du} < L^*\underset{\sim}{\lambda},\underset{\sim}{z} > = < \frac{dL^*}{du} \underset{\sim}{\lambda},\underset{\sim}{z} > + < L^*\underset{\sim}{\lambda}, \frac{d\underset{\sim}{z}}{d\underset{\sim}{u}} > = 0 \quad .$$

Therefore,

$$3.4) \quad \frac{d\psi_o}{d\underset{\sim}{u}} = - < \left(\frac{dL^*}{d\underset{\sim}{u}}\right)\underset{\sim}{\lambda}, \underset{\sim}{z} > + \frac{\partial\psi_o(\underset{\sim}{z},\underset{\sim}{u})}{\partial\underset{\sim}{u}} \quad .$$

Somehow, we managed to describe the sensitivity of the cost functional with respect to the design without knowing sensitivity of the state function (that is solution of our differential equations of state) with respect to the design. The case discussed above was very simple. Only the cost functional was involved in formulation of the functional $a(\underset{\sim}{z},\underset{\sim}{\lambda})$ instead of a vector ψ_α, $\alpha = 0,1,2..r$, which would incorporate constraints. The inhomogenous terms Q(z), q(z) were assumed to be independent of the design. All kinds of smoothness was tacitly assumed. For all that the equation (3.4) has remarkable simplicity. To compute $\frac{d\psi_o}{d\underset{\sim}{u}}$ at $\underset{\sim}{u} = \underset{\sim}{u}_o$ we need only to know the vector $\underset{\sim}{\lambda}(\underset{\sim}{x})$, i.e., to solve the system of adjoint equations with the simplest (generally zero) conditions assigned to the boundary $\partial\Omega$ and to initial or finite states of the system. The sensitivity $\frac{dL^*}{d\underset{\sim}{u}}$ is easily computed directly and so is $\frac{\partial\psi_o}{\partial\underset{\sim}{u}}$.

In essence, this is the main idea of the adjoint variable "trick". In general, if $\underset{\sim}{\psi}$ is a vector as given in (3.1) and $\underset{\sim}{\psi}$ lives in a Banach space B_1, while $\underset{\sim}{u}$ lies in a Banach space B_2, $\frac{\partial\psi}{\partial\underset{\sim}{u}}$ is an operator mapping B_2 into B_1 and the product $< \; >$ is not the usual inner product, but denotes a bilinear operator valued function, rather than an inner product in some Hilbert space. Similarly, one has to interpret correctly the meaning of $\frac{\partial \underset{\sim}{z}}{\partial \underset{\sim}{u}}$, $\frac{\partial L^*}{\partial \underset{\sim}{u}}$, and so on. These are technical details which could invalidate the entire analysis if they are not checked and if the mathematical background is not verified. But careful attention to such detail in a very general case would obscure the simplicity of the main idea. The engineers familiar with Pontryagin's approach to control theory may immediately recognize the similarity between this treatment of the adjoint operator and the duality for the Pontryagin's Hamiltonian function.

I will conclude by conjecturing that bifurcation phenomena for cost functional in the optimal design for dynamical systems will display the familiar switching pattern of optimal control problems that are usually derived by the study of the Hamiltonian. So far, a complete analogy has eluded the researchers in this field. No paper discussing sensitivity and optimization of design would be complete without at least a mention of some numerical "tricks of the trade" used in handling of "local" constraints or some "local properties of the cost functional.

Roughly speaking, global constraints are "good" and fit well with variational arguments and local constraints are "bad". In [45] Banichuk offers many suggestions for conversion of $L^\infty(\cdot)$ constraints to $L^P \geq 1$, trying to deal with this problem.

However, a theoretical justification of some steps involved in such replacement has not been given to the best of our knowledge.

In a 1982 article Gilbert Strang [48] shed some light on this difficult theoretical problem of approximating L^1 and/or L^∞ norms for vector fields. Strang's discussion is restricted to two dimensions and the differential operator is the equivalent of the Laplace Operator (That is, optimization in the L^2 norm yields the Laplace equation). The author admits that, in general, the L^∞ optimization problems remain open at the present time. As usual, W. Prager cleverly by-passed the problem of non-uniform convergence, relying on purely engineering arguments in the discussion of L^∞ optimization of structures and in particular in the problem of pure torsion. This aspect of L^∞ optimization is discussed also in a 1982 paper of G. Strang and R. Kohn published in a collection of articles on the Finite Elements and Applications by the Academic Press [49].

Other computational tricks, such as converting local constraints to global constraints can be found in several papers such as [33], [34]. For example, the maximum stress condition $\frac{P}{A} < \sigma_{max}$, which is a local condition, can be replaced by

$$\int_\Omega \left| \; |\sigma_{max} - \frac{P}{A}| - (\sigma_{max} - \frac{P}{A}) \; \right| d\underset{\sim}{x} = 0 \quad .$$

This simple idea can be pursued for more complicated local constraints, as was indicated in the papers listed above. Related ideas originated by W. Prager may be found in [50], and in [51].

4. VARIATION OF THE DOMAIN PROBLEM, AND DOMAIN SENSITIVITY

4.0 A general discussion.

Around 1979 a technique called the "speed method" was developed at the University of Nice, with J. Cea predominantly involved in this development.

The material derivative (in the engineering terminology) is used to generate a family of transformations. Speed of change in the shape is used to define the sensitivity. Since the material derivative is the Lie derivative with respect to an exterior form it is a natural consequence of the success of this method that group theoretic techniques should make some inroads into design optimization and eventually into Computer Assisted Design algorithms.

Let us briefly review the possible future foundations of this approach.

4.1 Change of Shape

We can introduce a one parameter family of maps $\Omega \to \Omega_\tau$. Let us suppose that a shape $\Omega \subset R^n$ is dynamically deformed with the family of shapes Ω_τ uniquely defined for all values of $0 \leq \tau \leq 1$ and the map $T_\tau : \Omega \to \Omega_\tau$ is a continuous homeomorphism. Each point x Ω is continuously moving along a simple arc $T_\tau : x \to x_\tau$, where $T_{\tau=0}=I$, is the identity map on Ω.

We presume that T_τ defines a strongly continuous family of operators which form a semigroup under composition:

(1.1) $$T_\mu \cdot T_\tau (\Omega) = T_\mu(\Omega_\tau) = T(\Omega_{\mu+\tau}) = T_{\mu+\tau}(\Omega).$$

The infinitesimal generator of this semigroup is given by

$$\lim_{\tau\to 0} \left[\frac{T_\tau - I}{\tau} \right] = t_o \quad .$$

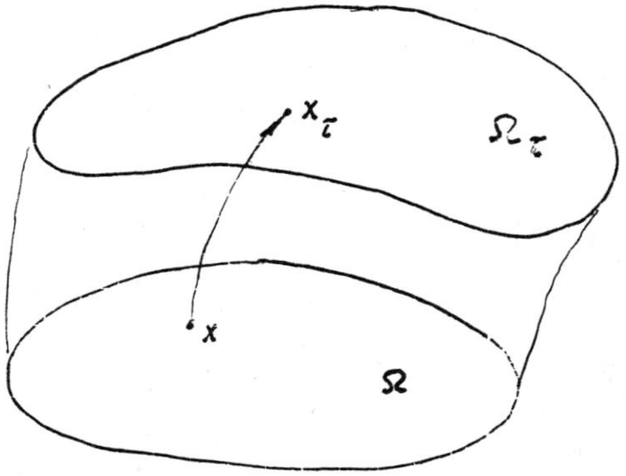

Figure 3

$\dot{T}_0$ defines physically the initial velocity operator. Specifically, $\dot{T}_0 x_0$ =

(1.2) $$\lim_{\tau\to 0} \frac{1}{\tau} [T_\tau(x_0) - x_0] = \lim_{\tau\to 0} \frac{1}{\tau} [x_\tau - x_0] = \frac{dx_\tau}{d\tau}\bigg|_{\tau=0} .$$

In a sufficiently small neighbourhood of zero one can estimate the deformation at x by writing

(1.3) $$x_\tau = x_0 + \tau\cdot\dot{T}_0 x_0 + r(\tau,x_0) .$$

where $r(\tau,x_0)$ is the remainder obeying the limit relation:

(1.3[a]) $$\lim_{\tau\to 0} \frac{1}{\tau} r(\tau,x_0) = 0 .$$

The function $\dot{T}_0 x_0$ defines the sensitivity of the shape Ω to the deformation process described by the operator family T_τ computed at $x_0 \in \Omega$.
For an arbitrary function, or functional $\Phi(x)$, $x \in \Omega$, $\Phi : \Omega \to \mathbb{R}$ we define the material (Lie) derivative along the action of the semigroup of transformation T_τ to be

(1.4) $$\lim_{\tau\to 0} \frac{\Phi_\tau\,(x + \tau\,\dot{T}_0 x_0) - \Phi\,(x)}{\tau} = \dot{\Phi}\,(x) ,$$

where $\Phi_\tau\,(y) = \Phi\,(y_\tau)$.

Similarly for an arbitrary function $z(x)$, $x \in \Omega$ we define

(1.5) $$\dot{z} = \lim_{\tau\to 0} \frac{1}{\tau} \{z_\tau\,(x + \tau\dot{T}_0 x) - z(x)\} .$$

In problems involving continuum mechanics or other continuous phenomena pointwise definitions are inappropriate and our definition should be corrected to read $\dot{z}(x)$ is defined almost everywhere in the $H_0^m\,(\Omega)$ sense by the relation

(1.5[a]) $$\lim_{\tau\to 0} \left\| \frac{1}{\tau} \{z_\tau(x + \tau\dot{T}_0 x) - z(x)\} - \dot{z}(x) \right\|_{H_0^m(\Omega)} = 0 .$$

$\dot{z}$ is frequently called the material derivative.
Here $H_0^m(\Omega)$ is the appropriate Sobolev space assigned to our problem. By Sobolev imbedding lemma if $2m > n$, $H_0^m(\Omega)$ is a subspace of $C(\Omega)$ (the class of continuous functions) and pointwise definitions make sense. Otherwise only $L^2(\Omega)$ averages make physical and mathematical sense and all concepts defined above must be interpreted in the sense of $H_0^m(\Omega)$ average quantities. As the shape Ω is transformed,

the basic state equations given below

$$(1.6)\quad \left.\begin{array}{ll} Az = f & \text{in } \Omega , \\ z \equiv 0 & \text{on } \partial\Omega \end{array}\right\}$$

are conserved.

That is

$$(1.6^a)\quad \begin{array}{ll} Az_\tau = f & \text{in } \Omega_\tau \\ z_\tau = 0 & \text{on } \partial\Omega_\tau \ . \end{array}$$

The system (1.6^a) could be intepreted as a weak equation in $H_0^m(\Omega)$. The variational form of equation (1.6^a) is obtained in Ω_τ:

$$(1.7)\quad a_\tau(z_\tau,\bar{z}_\tau) = \ell_\tau(\bar{z}_\tau) \quad \text{in } \Omega_\tau \quad z_\tau \in Z_\tau ,$$

where

$$(1.8)\quad a_\tau(z_\tau,\bar{z}_\tau) = (Az_\tau, \bar{z}_\tau)_{\Omega_\tau} ,$$

$$(1.9)\quad \ell_\tau(\bar{z}_\tau) = (f, \bar{z}_\tau)_{\Omega_\tau} .$$

The bilinear form $a(z_\tau,\bar{z}_\tau)$ is regarded as the Friedrichs form, that is, both the domain and range of the operator A has been changed. Hence, it is not the same operator, but the extension of A. However, we use the same symbol for the operator A as before, since no confusion can arise.

4.2 The material derivative.

Let J be the Jacobian matrix associated with the transformation T_τ.

$$(2.1)\quad J_\tau = \frac{\partial T_\tau}{\partial x} = I + \tau \frac{\partial \dagger}{\partial x} + r(\tau^2) \quad , \text{ where } \left(\frac{\partial T}{\partial x} \equiv \left[\frac{\partial T_{\tau\hat{i}}}{\partial x_{\hat{j}}} \right]\right) ,$$

$$J_{\tau=0} = I .$$

By assumption the map T_τ is a homeomorphism and the matrix J_τ is nonsingular for the considered values of τ. Hence J_τ^{-1} exists.
A simple computation shows that

$$(2.2^a)\quad \lim_{\tau\to 0} \left| J_\tau \right| = \operatorname{div} (\dagger_o) ,$$

and taking the limit $0 = \lim_{\tau\to 0} \frac{d}{d\tau} \left|JJ^{-1}\right|$, one obtains

$$(2.2^b) \quad \lim_{\tau\to 0} \left| J_\tau - 1 \right| = \mathrm{div}\,(\dot{T}_0) \quad .$$

We define the material derivative of a functional $\Phi(f_\tau) = \int_{\Omega_\tau} f_\tau(x_\tau)\, d\Omega_\tau$ computed at $\tau = 0$ to be the functional.

$$(2.3) \quad \Phi_0' = \frac{d}{d\tau} \int_\Omega [f_\tau(x + \tau\dot{T}_0) \cdot J_\tau]\, d\Omega \,\Big|_{\tau = 0}$$

$$= \int_\Omega [f'(x) + (\nabla f(x), \dot{T}_0) + (f(x)\, \mathrm{div}\, \dot{T}_0)]\, d\Omega$$

$$= \int_\Omega [f'(x) + \mathrm{div}(\dot{T}_0 f)]\, d\Omega \; .$$

f' denotes the partial derivative

$$f'(x) = \lim_{\tau\to 0} \frac{f_\tau(x) - f(x)}{\tau} \quad .$$

Using divergence theorem, one can transform equation (2.3) to the form

$$(2.4) \quad \Phi_0' = \int_\Omega f'(x)\, d\Omega + \int_{\partial\Omega} f(x)(\dot{T}_0 \cdot \vec{n})\, d(\partial\Omega).$$

where $\vec{n}$ is the unit vector normal to $\partial\Omega$.
Sufficient regularity conditions have to be assumed concerning $\partial\Omega$ for the second (boundary) integral to make sense.
Equation (2.3) defining the material derivative can be recognized as another definition of a Lie derivative for an exterior form of order zero (i.e. for a function).
The equation (2.4) is crucial in deriving some sensitivity results.
Functions of the form

$$(2.5) \quad \Phi = \int_{\Omega_\tau} f(z_\tau, \mathrm{grad}\, z_\tau)\, d\Omega_\tau$$

have material derivative given by:

$$(2.6) \quad \Phi_0' = \int_\Omega (f(z) \cdot z' + f_{\nabla z}\, \nabla z')\, d\Omega$$

$$+ \int_{\partial\Omega} f \cdot (\dot{T}_0 \cdot \vec{n})\, d\,(\partial\Omega) \quad .$$

We denote by $\dot{z}$ the form

(2.7) $\dot{z} = z' + \dot{T}_0 \cdot \nabla z$,

where as before

$$z' = \lim_{\tau \to 0} \left(z_\tau \frac{(x) - z_0(x)}{\tau} \right) .$$

Then

(2.8) $$\Phi_0' = \int_\Omega [f(z)\,\dot{z} + f_{\nabla z}\,\nabla\dot{z} - f_z \cdot (\dot{T}_0\,\nabla z) - f_{\nabla z}\,\nabla\,(\dot{T}_0\,\nabla z)]\,d\Omega + \int_{\partial\Omega} f\,(\dot{T}_0 \cdot n)\,d(\partial\Omega) .$$

where $f_{\nabla z} = \{ \frac{\partial f}{\partial z_1}, \frac{\partial f}{\partial z_2} , \frac{\partial f}{\partial z_3} \}$.

In a special case the functional Φ is interpreted as a bilinear functional. For example,

(2.9) $$a(z, \lambda) = \int_\Omega [(f_z, \lambda) + (f_{\nabla z}, \nabla\lambda)]\,d\Omega .$$

If we identify λ with $\dot{z}$, (2.9) becomes

(2.9[a]) $$a(z,\dot{z}) = \int_\Omega [f_z\dot{z} + f_{\nabla z}\,\nabla\dot{z}]\,d\Omega .$$

However, $a(z,\lambda)$ may be simultaneously identified with a bilinear form naturally arising in physical consideration, such as, for example, the virtual work performed on a structural system by the external loads, thus providing us with the study of sensitivity of certain energy terms to the changes in the domain.

4.3 Applications to design practice.

We analyze the beam design optimizing the total weight. The cost functional Φ is given by

$$\Phi = \int_0^\ell \rho\,A(x)\,dx .$$

The material derivative of Φ is given by

$$\Phi' = \int_0^\ell (\rho A)'\,dx + \int_{\partial\Omega} (\rho A)(\dot{T}_0 \cdot n)\,ds$$

$$= \int_0^{\ell} (\rho A)' \, dx + \rho A \, \dot{T}_0(\ell) - \rho A \, T_0(0).$$

$$= \rho \frac{d}{d\tau} \int_0^{\ell} (A_\tau(x + \tau \dot{T}_0) \cdot J_\tau) \, dx \Big|_{\tau = 0} + \rho A \dot{T}_0(\ell) ,$$

since we can vary the length by keeping one end fixed, with no apparent loss of generality.

Φ cannot be determined unless we first establish the mapping or the class of admissible mappings $\dot{T}_\tau : \Omega \to \Omega_\tau$. Such mapping is easily established by fairly "conventional" (i.e. by now 5 - year old or older techniques) following either direct methods such as given in [56] or more sophisticated methods introduced in [57] and further expanded by Haug and associates in a series of articles. (See for example [58], [59]).

To illustrate this point we offer a fairly easy computation of sensitivity of the complience functional for an elastic beam. We assume the Euler-Bernoulli linear beam theory with a distributed load $q(x) \in L_2[0,\ell]$ and the specific weight of the beam $g\rho A(x)$ contributing to the applied load.

The natural setting for this problem is the Sobolev space in which the inner product is $\langle f,g \rangle = \int_\Omega (EI(x) \, f_{xx} g_{xx}) dx$.

We write the basic equation of equilibrium for an elastic beam

$$(3.2) \quad (EI(x)W_{xx})_{xx} = q(x) + g\rho A(x) ,$$

equating the second derivative of the bending moment with the load applied (q(x)) plus the weight of the beam regarded as additional loading. We assume constant geometry and a relation $I(x) = \phi(A(x))$, which is associated with certain additional geometric assumptions of our model. $\phi(A(x))$ is assumed to be a differentiable function of A(x).

As usual, ρ is the material density, E - the Young modulus, A(x) - the cross-sectional area, I(x) - the moment of inertia of the cross-sectional area about the neutral axis, g is the earth's gravitational constant.

The potential energy binomial form is given by

$$(3.3) \quad a(w, \lambda) - b(w, \lambda) =$$

$$\int_0^{\ell} [E\phi(A(x)) \, w_{xx} \, \lambda_{xx}] \, dx -$$

$$\int_0^{\ell} (q(x) \, \lambda + g\rho A(x) \, \lambda \,) \, dx \; .$$

The load $\underset{\sim}{q}(x)$ may depend on w, and for sake of greater generality we shall assume that

$$q(x) = \hat{q}\ (w(x),\ x)\ .$$

Let the cost functional be given by the complience, i.e.

$$\Phi(w) = \int_0^{\ell}[(q(w(x),\ x) + g\rho A(x))w(x)]dx$$

The corresponding bilinear form is given by

$$\hat{\Phi}(w,\lambda) = \int_0^{\ell}(q\ (w(x),\ x) + \rho g A(x))\lambda(x)]dx\ .$$

The change of shape of the beam consists of changing the parameter A(x) and the length ℓ . The variation of A(x) in a singular case (torsion only) is given in the article of Hou, Haug and Benedict in this issue. Let us vary the length only. The variation of the domain is fairly straight-forward in this case.

The variation of the total weight is given by the material derivative

$$(3.4)\quad [\int_0^{\ell}\rho\ A(x)\ dx\ ]' = \rho A\ \dot{T}\Big|_0^{\ell} = \rho A(\ell)\dot{T}(\ell) - A(0)\dot{T}(0)\ .$$

The material derivative of the potential energy is given by

$$(3.5)\quad U' = \int_0^{\ell}[EI(A)\ \lambda_{xx})_{xx}\ (\lambda_x T_\tau) - (q\lambda_x\dot{T}_\tau)]\ dx$$

$$+ EI(A)\ \lambda_{xx}(\lambda_x\dot{T}_\tau)_x\Big|_0^{\ell} - (EI(A)\lambda_{xx})_x \cdot (\lambda_x\dot{T}_\tau)\Big|_0^{\ell} + q(x)\lambda - EI(A)(\lambda_{xx})^2\ \dot{T}_\tau\Big|_0^{\ell}\ .$$

Identifying λ and w and simplifying this expression by assuming that beam is clamped at both ends, one obtains an extremely simple sensitivity formula in terms of the boundary transformation rate

$$(3.6)\quad U' = -\ EI(A)w_{xx}{}^2\ \dot{T}_\tau\Big|_0^{\ell}$$

This conclusion is fairly obvious if intuitive physical arguments are offered. However, this is not the case if one considers even simple cases of plate or shell designs.

More challenging numerical computations that are using this technique will be given in the monograph of Haug, Choi and Komkov [51], to appear in 1985 (Academic Press, New York).

However, even such simple one-dimensional analysis may offer nontrivial insight into the design alteration procedures.

Example. We consider one of the extensively researched class of problems in structural mechanics - that of optimization, sensitivity and differentiability of the natural frequencies. If we accept the Euler - Bernoulli model for a vibrating beam, the sensitivity of a simple eigenvalue corresonding to the fundamental natural frequency is given by

$$4.0) \quad \zeta' = -2 \int_0^{\ell} \{EI(A(x))w_{xx}\,(w_x \dot{T}_\tau)_{xx} + \zeta\rho A(x)w\,(w_x \dot{T}_\tau)\}\,dx + [EI(A(x))(w_{xx})^2 - \zeta\rho A(x)w^2]\dot{T}_\tau \Big|_0^{\ell} .$$

If we assume the clamped - clamped support conditions at $x = 0$ and $x = \ell$, and recall that $w(x)$ is an eigenvector corresponding to the simple eigenvalue ζ we can derive the simplified sensitivity formula

$$(4.1) \quad \zeta' = -\,EI(A(x))\,(w_{xx})^2\,\dot{T}_\tau \Big|_0^{\ell} ,$$

indicating the effects on changes of the natural frequency caused by the changes in length and in the bending moment $EIw_{xx} = M(x)$ (therefore in w_{xx}) at both ends points. It is clear that ζ' is a quadratic function of the "symmetric moment' $\mathcal{M}(x) = (EI\,)^{\frac{1}{2}}\,w_{xx}$ at the end points of the beam. To decrease the natural frequency one should move outward the end point of the beam at which $|\,(EI)^{\frac{1}{2}}\,w_{xx}|$ is bigger. The complex problem of "crossover" of eigenvalues (as discussed resulting from an iterative application of such iterative procedure (for a constant total weight)) is deliberately avoided here.

5. INVARIANT VARIATIONAL FORMULATION OF OPTIMAL DESIGN, AND DESIGN SENSITIVITY OF A COLUMN, OR A VIBRATING BEAM.

We consider the design of an elastic beam obeying the usual Euler - Bernoulli linear hypothesis, obeying the equation

$$5.1) \quad (EI(x)y'')'' = q(x) ,$$

where E is a positive constant (Young's modulus) ,
$I(x)$ is a positive function (moment of inertia of the cross-sectional area about the neutral axis of bending) and $q(x)$ is the applied load. The displacement func-

tion y(x) is an element of $H^2[0,\ell]$ (sobolev) space, $q(x) \in L_2[0,\ell]$. In the variational formulation one seeks to find a stationary behavior of the bilinear functional

(5.2) $V = < EI(x)\, y'', \eta'' > - < q, \eta >$

where <,> is the usual $L_2[0,\ell]$ inner product.
We can consider equation (5.1) as a constraint applied to the problem of minimizing the functional

$$\Phi(y(x), A(x)) \to \min.$$

A special case is the problem of minimizing the weight

$$\Phi'(A) = \int_0^\ell \rho A(x)\, dx \;, \; \rho = \text{constant}$$

which is equivalent to the minimization problem

$$\Phi(A(x)) = \int_0^\ell A(x)\, dx \to \min.$$

A constraint applied to the problem may be the maximum value of stress, or the minimum value of natural frequency, or the maximum value of displacement. Let us suppose that we wish to restrain the minimum value of the fundamental (natural) frequency, i.e. we assign a constraint to the Raybeigh quotient

$$w^2 = \min_{y \in Y} \{\int_0^\ell EI(A(x)(y'')^2\, dx \,/\int_0^\ell (y')^2 dx\}$$

$$\geq \kappa \,,$$

where κ is given and y is chosen in the space Y of admissible displacement functions. Clearly, Y is a subset of the Sobolev space $H_0^2[0,\ell]$, if $y(0) = y(\ell) = 0$.

The problem can be restated by requiring

$$\tfrac{1}{2} \int_0^\ell (y')^2 dx = 1$$

and minimizimg

$$\tfrac{1}{2} \int_0^\ell EI(A(x))(y'')^2\, dx \;.$$

The entire optimization of shape problem is reduced to the study of the stationary points of the functional

$$(5.3)\quad \lambda = \lambda(A(x), y(x)) = \int_0^\ell A(x)\,dx - \tfrac{1}{2}[\beta \int_0^\ell EI(A(x))(y'')^2\,dx - \kappa] - \tfrac{1}{2}\gamma\,[\int_0^\ell (y')^2\,dx - 1],$$

where β, γ are Lagrangian multipliers.

The functional λ is invariant under the action of the group G_k that is mapping x into $\bar{x}$, $y \in H^2[0,\ell]$ into $\bar{y} \in H^2[0,\ell]$, $E \to E$, $A \to A(x) \in H_0^1[0,\ell]$.

We assume the relation $\bar{E} = E(1 + \varepsilon_k)$ to be the only one considered.

The functional λ is invariant if for each $i = 1,2,..k$, the following equation is satisfied

$$\frac{\partial\lambda}{\partial x}\tau_i + \frac{\partial\lambda}{\partial A}\nu_i + \frac{\partial\lambda}{\partial y}\xi_i + \frac{\partial\lambda}{\partial y_x}\left(\frac{d\xi_i}{dx} - y_x\frac{d\tau_i}{dx}\right) + \frac{\partial\lambda}{\partial y_{xx}}\left(\frac{d^2\xi_i}{dx^2} - 2y_{xx}\frac{d\tau_i}{dx} - y_x\frac{d^2\tau_i}{dx^2}\right)$$

$$+ \frac{\partial\lambda}{\partial E}\mu_i + \lambda\frac{d\tau_i}{dx} = 0\,, \qquad \text{where}$$

$\frac{d}{dx}$ denotes the total derivative,

$$\text{i.e.}\quad \frac{d\xi_i}{dx} = \frac{\partial\xi_i}{\partial x} + \frac{\partial\xi_i}{\partial y}y_x + \frac{\partial\xi_i}{\partial y_x}y_{xx} + \frac{\partial\xi_i}{\partial y_{xx}}y_{xxx} + \frac{\partial\xi_i}{\partial A}\frac{dA}{dx}\,,$$

etc.

ε_k will denote the parameters of G_k, $\underset{\sim}{\varepsilon}$ the vector $\underset{\sim}{\varepsilon} = \{\varepsilon_1, \varepsilon_2, ..\varepsilon_k\}$, μ_i, η_i, τ_i, ξ_i are the infinitesimal generators of G_k.

Let $\bar{x} = \phi(x, y, A)$, $\bar{y} = \Psi(x, y, A)$

$\bar{A} = \eta(x, y, A)$, $\bar{E} = \mu(E)$.

Then

$$\tau_i = \frac{\partial\phi}{\partial\varepsilon_i}\Big|_{\underset{\sim}{\varepsilon}=0} \qquad \xi_i = \frac{\partial\Psi}{\partial\varepsilon_i}\Big|_{\underset{\sim}{\varepsilon}=0}$$

$$\nu_i = \frac{\partial\eta}{\partial\varepsilon_i}\Big|_{\underset{\sim}{\varepsilon}=0} \qquad \mu_i = \frac{\partial\bar{E}}{\partial\varepsilon_i}\Big|_{\underset{\sim}{\varepsilon}=0} \quad \left(= 1 \text{ if } i = k \text{ and zero otherwise}\right)$$

We notice that nontrivial generators μ_i will arise in some imaginative uses of composite materials.

Let us consider a six parameter group of transformations G_6 :

$\bar{x} = x + \varepsilon_1 x\ (\ell - x)$

$\bar{y} = y + \varepsilon_e x + \varepsilon_3 y$

$\bar{A} = A(1 + \varepsilon_4) + \varepsilon_5 y$

$\bar{E} = E(1 + \varepsilon_6)$

Then $\tau_1 = x(\ell - x)$, $\tau_2 = \tau_3 = \tau_4 = \tau_5 = \tau_6 = 0$,

$\xi_1 = 0,\ \xi_2 = x,\ \xi_3 = y,\ \xi_4 = \xi_5 = \xi_6 = 0$,

$\nu_1 = \nu_2 = \nu_3 = 0,\ \nu_4 = A,\ \nu_5 = y,\ \nu_6 = 0$,

$\mu_1 = \mu_2 = \mu_3 = \mu_4 = \mu_5 = \mu = 0,\ \mu_6 = E$.

The Noether equations which formulate a necessary condition for the invariance of λ are:

a) $\frac{\partial\lambda}{\partial x} x(\ell - x) + \frac{\partial\lambda}{\partial y_x} [\ y_x(2x - \ell)\ ] + \frac{\partial\lambda}{\partial y_{xx}} [2y_{xx}\ (2x - \ell) - 2y_x] + \lambda\ (-2x + \ell) = 0$

b) $x \frac{\partial\lambda}{\partial y} + \frac{\partial\lambda}{\partial y_x} = 0$

c) $\frac{\partial\lambda}{\partial y} y + \frac{\partial\lambda}{\partial y_x} y_x + \frac{\partial\lambda}{\partial y_{xx}} y_{xxx} = 0$

d) $A \frac{\partial\lambda}{\partial A} = 0$

e) $y \frac{\partial\lambda}{\partial A} = 0$

f) $E \frac{\partial\lambda}{\partial E} = 0$.

Conditions d) and e) can be combined into a single condition

(d') $(y(x)A(x) \frac{\partial\lambda}{\partial A}) = 0$

Condition (f) is equivalent to

(f') $\frac{\partial\lambda}{\partial E} = 0$

We observe that equation (f') arises as a necessary condition for stationary behavior of λ for any transformation of coordinates which implies replacing E by $\bar{E}$

but other coordinate changes are not affected by the value of E. However in linear (that is Euler-Bernoulli) theory this stationary behavior condition is impossible to satisfy.

The sensitivity of λ as a function of E is determined by the infinitesimal generator $E \frac{\partial\lambda}{\partial E} = M(x)$, and clearly, only the trivial condition $M(x) = EI(x)\, y'' \equiv 0$ is equivalent to optimality of λ. $M(x)$ is the exact bending moment distribution caused by the applied load. Here, again, we discover some obvious limitations of the linear theory and of using only homogeneous materials in our design (i.e. E is independent of x or A).

Some invariants associated with the group G_6 are found by recalling a standard procedure of Noether theory. See [52] for details of the basic theory and [53], [54] for specific computations.

For example, the following invariant associated with G_6 group of transformations is easily established.

$$C = \lambda\tau_1 + \left(\frac{\partial\lambda}{\partial y_x} - \frac{d}{dx} \frac{\partial\lambda}{\partial y_{xx}} \right) (\xi_1 - y_x\tau_1)$$

$$+ \frac{\partial\lambda}{\partial y_{xx}} \frac{d}{dx} (\xi_1 - y_x \tau_1) =$$

$$\lambda x(\ell - x) - \left(\frac{\partial\lambda}{\partial y_x}\right) y_x \cdot x (\ell - x) + \frac{d}{dx} \left(\frac{\partial\lambda}{\partial y_{xx}}\right) y_x\, x(\ell - x)$$

$$= x (\ell - x) \left\{ \lambda - \left[y_x \left(\frac{\partial\lambda}{\partial y_x} - \frac{d}{dx} \left(\frac{\partial\lambda}{\partial y_{xx}}\right)\right)\right] \right\}.$$

Since this quantity is invariant along the entire length of the column, we easily derive the value of C, namely

$$C \equiv 0 \text{ for all } x \in [0, \ell].$$

Other invariants are similarly derived. As in other problems of physics, it appears that the clever choice of the "right" group of transformations will significantly lighten the numerical procedures of improving an engineering design. This aspect of theoretical research opens new possibilities of study of engineering processes by associating certain "natural" groups of transformations to design problems in applied mechanics in the same manner in which the Lorentz group "naturally" fits the classical equations of Maxwell in the electromagnetic field theory.

This approach is in its infancy, and only first steps have been taken to develop some systematic procedures.

6. Other comments.

N.V. Banichuk used a purely formal application of a sensitivity formula similar to formulas 2.2-2.4. in the optimization shape resulting in maximal rigidity for membranes and elastic strips (Chapter 1 [45]). A completely different approach to the problem of domain optimization and domain sensitivity was attempted in [55].

So far, it has not progressed beyond basic definitions and simplest cases that are easily solved by other techniques.

As a final comment I wish to add that topics and future directions that are discussed in this article represent only the taste and the interests of the author. Many important aspects and modern developments of sensitivity theory and of its applications to mechanical or civil engineering have been completely neglected. For example, the entire recent development of optimization and sensitivity for mechanical systems based on game-theoretic principles, has been entirely by-passed here. This particular approach to min-max problems has been actively pursued in the research papers of F.L. Chernous'ko.(See [60], [61], or [62]) and is outlined in the F.L. Chernous'ko and A.A. Melikjan monograph [63]. A related numerical technique called remodeling has been pursued by John Taylor (See [65]), and [66]). Numerical techniques were not mentioned except where they generated important theoretical principles.

For a review of recent literature see [64], and the addendum to bibliography given in the English translation of Banichuk's monograph [45].

Other topics which have been omitted here include treatment of constraints by penalty method that is strongly related to similar techniques in control theory (as pioneered by A.V. Balakrishnan), the related free boundary problems such as contact problems in elasticity, and a host of related numerical techniques, such as various gradient projection techniques including Miele's and Uzawa's algorithms. A sampling of algorithmic procedures related to this approach may be found in J. Cea's monograph [67]. For original ideas related to the numerical applications of "speed" method see the original article of J. Hadamard [68].

We have also deliberately ignored an extremely important topic. We have discussed only conservative systems, i.e. the systems represented by symmetric operators. For example, no dissipation was considered. Some discussion of non-selfadjoint operators in structural vibration problems is offered in the paper of P. Pedersen [47] (in this issue). Pursuing the Noether ideas of reference [52] a group theoretic approach to certain classes of non-selfadjoint problems of applied mechanics was given by the author in [53], but at the present time the general concepts introduced in that work have not been applied to either design or sensitivity theory, and no numerical techniques have been suggested that are based on such group theoretic principles which were specifically directed at sensitivity or optimization of engineering design.An application to electrodynamics is given in [69],to hydrodynamics in [70] and to linearalized,finite elasticity in [71]. Some promising generalizations that are directly applicable to continuum mechanics appeared in the work of B.Vujanović [72].

References

1. Leonhardi Euleri Opera Omnia, Vol. X, ser. secundae, Society for Natural Sciences of Switzerland, 1960, in particular the section of C. Truesdell's historical notes, p. 1638-1788, on the rational mechanics of flexible or elastic bodies.

2. J.L. Lagrange Sur la figure des colonnes, Miscellanea Taurinensia, Vol. V, 1970, (see p. 123-125).

3. J.B. Keller, The shape of the strongest column, Archives of Rational Mechanics and Analysis, Vol. 5 (1960), p. 275-285.

4. I. Tadjbakhsh and J.B. Keller, Strongest columns and isoperimetric inequalities for eigenvalues, J. of Applied Mechanics, Vol. 9, (1962), p. 159-164.

5. N. Olhoff, Optimal design against structural vibration and instability, Ph.D. Thesis, Technical University of Denmark, Dept. of Solid Mechanics, Lyngby, Denmark, November 1978.

6. E.J. Haug, U.S. Army Material Command Pamphlet, AMC 706-902 (1972-73), Engineering Design Handbook.

7. B.M.E. DeSilva, Applications of Pontryagin's principle to a minimum weight design problem, Journal of Basic Engineering, ASME, Vol. 1, #92, (1970), p. 245-250.

8. J.L.P. Armand, Applications of Optimal Control Theory of Systems with Distributed Parameters to Problems of Structural Optimization (in Russian) Mir, Moscow, 1977.

9. Vadim Komkov and N. Coleman, Optimality of design and sensitivity analysis of beam theory, Int. J. Control, Vol. 18, #4 (1973), p. 731-740.

10. Vadim Komkov, Simultaneous Control and Optimization for Elastic Systems, Proceedings of International Conference on Applications of Distributed System Theory to the Control of Large Space Structures, J.P.L. Pasadena, California, July 1982, N.A.S.A., 1983.

11. J.B. Keller and F.I. Niordson, The tallest column, J. of Math. and Mechanics, Vol. 16, (1966), p. 433-466.

12. E. Masur, Singular problems of optimal design, in Optimization of Distributed Parameter Structures NATO Symposium, Iowa City, IA, E.J. Haug and J. Cea Editors, Noordhoff and Sijthoff Publishers, Holland (1980), p. 200-218.

13. K.K. Choi and E.J. Haug, Optimization of Structures with Repeated Eigenvalues, Ibid, p. 219-277.

14. N. Olhoff and J. Taylor, Designing Constinuous Columns for Minimal Cost of Material and of Interior Supports, J. of Structural Mechanics, Vol. 6, (1978), p. 367-382.

15. Z. Mróz and G.I.N. Rozvany, Optimal Design of Structures with Variable Support Conditions, J. Optimization Theory and Applications, Vol. 15, #1, (1975), p. 85-101.

16. E.F. Masur and Z. Mróz, Singular Solutions in Structural Optimization Problems, Proceedings IUTAM Sumposium. Springer Verlag, Berlin, 1975.

17. Emmy Noether, Invariante Variationsprobleme, Kgl. Gess. Nachrichte Göttingen, Math. Ph. K., 1918.

18. D.J. Logan, Invariant Variational Principles, Academic Press, New York, 1977.

19. Andrzej Trautman, Noether's equations and convservation laws, Commun. Math. Physics, 6, (1967), p. 248-261.

20. A.M. Arthurs, Complementary variational principles, Oxford Press, Oxford, 1968.

21. V. Komkov, Applications of Rall's theorem to classical elastodynamics, J. Math. Anal. Appl., 14, (1966), p. 511-521.

22. I.H. Ibrahimov, Invariant variational problems and conservation laws (comments on the theorem of E. Noether), Theoret. Mat. Fiz. 1, #3, (1969), p. 350-359.

23. V. Komkov, An embedding technique in problems of elastic stability, Z.A.M.M., 60, (1980), p. 503-507.

24. V. Komkov and C. Irwin, Proceedings A.M.S. meeting in New York, special session on sensitivity of functionals, April 1983.

25. V.G. Litvinov, Optimal Control Problem for the Fundamental Frequency of a Plate having a Variable Thickness, Vichesl. Mat. i Mat. Fiz. 19, #4 (1979), p. 866-877.

26. V.A. Krys'ko, The Optimal Control Problems for the Fundamental Frequency of Inhomogeneous Shells, Prikl. Mekh. Vol. 18, #4, (1982), p. 41-47. (Translated by Plenum Publ., New York, as Soviet Applied Mechanics)

27. A.P. Seyranian, Quasioptimal Solutions to Optimal Design Problems with Various Constraints, Soviet Applied Mechanics, Vol. 13, #6, (1977), p. 544-550.

(Also see references to A.P.Seyranian works in the contribution of P.Pederson in this issue.)

28. N.M. Gura and A.P. Seyranian, Optimum Circular Plate with Constraints on Rigidity and Fundamental Frequency of Vibration, M.T.T., Vol. 12, #1 (1977), p. 129-136.

29. A.P. Seyranian, Homogeneous Functions and Optimzation Problems, Int. J. Solids and Structures, Vol. 15 (1979), p. 749-759.

30. N. Olhoff and S.H. Rasmussen, On Single and Biomodal Optimum Buckling Loads for Clamped Columns, International Journal of Solids and Structures, Vol. 13, (1977), p. 605-614.

31. J.S. Arora and E.J. Haug, Optimum Structural Design Under Dynamic Loads, ASCE J. of Structural Division, Vol. 103, #ST.10, (1977), p. 2071-2074.

32. E.J. Haug and T.T. Feng, Optimization of Distributed Parameter Structures Under Dynamic Loads, Control and Dynamic Systems, Editor C.T. Leondes, Vol. 13, (1977), p. 207-246.

33. E.J. Haug, J.S. Arora, and T.T. Feng, Sensitivity Analysis and Optimization of Structures for Dynamical Response, ASME Journal of Mechanical Design, Vol. 100, (1978), p. 311-318.

34. M.H. Hsiao, E.J. Haug, and J.S. Arora, A State Space Method for Optimal Design of Vibration Isolators, ASME Journal of Mechanical Design, Vol. 101, (1979), p. 309-314.

35. E.J. Haug and J.S. Arora, Distributed Parameter Structural Optimization for Dynamic Response, in Optimization of Distributed Parameter Structures, editors E.J. Haug and J. Cea, Sijthoff and Nordhoff publishers, Netherlands 1980.

36. W. Velte and P. Villagio, Are the Optimum Problems of Structural Design Well Posed?, Archives of Rational Mechanics and Analysis, Vol. 78, # 3, (1982), p. 199-211.

37. V. Komkov, Application of Invariant Variational Principles to the Optimal Design of a Column, Z.A.M.M., Vol. 61, (1981), p. 75-80.

38. L.V. Ovsiannikov, Group-theoretic analysis of differential equations, Nauka, Moscow, 1978.

39. N.H. Ibrahimov, Lie-Becklund groups and conservation laws, D.A.N. USSR, 270, #1, (1976), p. 539-542.

40. W. Prager and R.T. Shields, Optimal Design of Multipurpose Structures, Int. J. Solids & Structures, Vol. 4, (1968), p. 469-475.

41. W. Prager and J. Taylor, Problems of Optimal Structural Designs, J. Applied Mech., Vol. 35, #1, (1968), p. 102-106.

[42] E.J. Haug and V. Komkov, Sensitivity Analysis in Distributed Parameter Mechanical Systems Optimization, JOTA, Vol. 23, #3, (1977), p. 445-464.

[43] J. Taylor and C.Y. Liu, On the Optimal Design of Columns, AIAA Journal, Vol. 6, #8, (1968), p. 1497-1502.

[44] E.J. Haug, Two Methods of Optimal Structural Design Development in Mechanics, Proceedings of 11th Midwestern Mechanics Conference, Development in Mechanics, 52 (1970), p. 847-860.

[45] N.V. Banichuk, Shape Optimization for Structural Systems, Nauka, Moscow, 1981, translated as: Problems and Methods of Optimal Structural Design, Plenum Press, New York, 1983.

[46] P. Pederson , Sensitivity Analysis for Non-selfadjoint problems, Preprint, presented at New York meeting A.M.S., April 1983. American Mathematical Society, Notices, March 1983.

[47] Pauli Pederson, Sensitivity Analysis for Non-Self Adjoint Problems, this issue.

[48] Gilbert Strang, L^1 and L^∞ Approximation of vector fields in the plane, in Non-linear PDE-s in Applied Science, U.S.A. - Japan Seminar, Tokoyo, Japan 1982. Published in Lecture Notes in Numerical and Applied Analysis, Volume 5 , (1982), Springer Verlag, Berlin and New York, 1982, pages 273-288.

[49] G. Strang and R. Kohn, optimal design of cylinders in shear, in the collection. The Mathematics of Finite Elements and Applications, Part IV, J. Whiteman editor, Academic Press, London and New York, 1982.

[50] G. I. N. Rozvany, Optimal Design of Flexural Systems, Pergammon Press, Oxford, 1976.

[51] W. Prager, Optimal design of statically determinate beams for a given deflection, Int. J. Mech. Science, Vol. 13.

[52] E. Noether, Invariante Variations Probleme, Nachv. Akad. Wiss. Gotingen, Math-Phys. Kl. II, 1981, 235-257.

[53] V. Komkov, A dual form of Noether's theorem with applications to continuum mechanics, J. Math. Anal. Appl., 75, #1, (1980), p. 251-269.

[54] A. Trautman, Noether's equations and conservation laws, Comm. Math. Phys. 6, (1967), p. 248-261.

[55] V. Komkov, The optimization of the domain problem I. Basic concepts, J. Math. Anal. Appl., Vol. 82, #2, (1981), p. 317-333.

[56] Banichuk, N.V., "Optimization of Elastic Bars in Torsion," Int. J. Solids and Structures, Vol. 12, 1976, pp. 275-286.

[57] Zolesio, J.P., "The Material Derivative (Or Speed) Method For Shape Optimization", Optimization of Distributed Parameter Structures (Eds. E.J. Haug and J. Cea), Sijthoff & Noordhoff Alphen aan den Rijn, Netherlands, 1980.

[58] Haug, E.J. and Rousselet, B., "Design Sensitivity Analysis of Shape Variations," Optimization of Distributed Parameter Structures, (Eds. E.J. Haug and J. Cea), Sijthoff and Noordhoff, Alphen aan den Rign, Netherlands, 1980.

[59] Zolesio, J.P. Identification de domaines par deformations, Thesis, Nice University, 1979.

[60] F.L. Chernous'ko, "Technique of Local Perturbations for Numerical Solution of Variational Problems," Zh. Vychisl. Mat. Fiz., Vol. 5, No. 4, pp. 749-754.

[61] F.L. Chernous'ko, "Certain Problems of Optimal Control With a Small Parameter," Prilk. Mat. Mekh., Vol. 32, No. 1, 1968, pp. 15-26.

[62] F.L. Chernous'ko, "Certain Optimal Shapes of Bifurcating Beams," Izv. Akad. Nauk SSSR, MTT, No. 3, 1979.

[63] F.L. Chernous'ko and A.A. Melikjan, Game-Theoretic Problems of Control and Search, Nauka, Moscow, 1978.

[64] E.J. Haug, "A Review of Distributed Parameter Structural Optimization Literature," Optimization of Distributed Parameter Structures(E.J. Haug and J. Cea,Ed.), Sijthoff & Noordhoff, Alphen aan den Rijn, Netherlands, 1981, pp. 3-68.

[65] J.E. Taylor and Martin P. Bendsøe, An Interpretation For Min-Max Structural Design Problems Including a Method For Relaxing Constraints, Int. J. Solids Structures, Vol. 20, No. 4, pp. 301-314, 1984.

[66] N. Olhoff and J.E. Taylor, On Optimal Structural Remodeling, J. Opt. Theory Applic. 27, 571-582 (1979).

[67] J. Cea, Lectures on Optimization - Theory and Algorithms, Springer-Verlag, 1978.

[68] J. Hadamard, "Mémoire sur le Probleme d'Analyse Relatif à l'Équilibre des Plaques Elastiques Encastrées (1908)", oeuvres de J. Hadamard, C.N.R.S., Paris, 1968.

[69] E. Bessel-Hagen, Über die Enhaltungssatze der Elektrodynamik,Math. Ann. 84 (1921), 258-276.

[70] S. Drobot and A. Rybarski, A variational principle in hydrodynamics, Arch. Rational Mech. Anal. 2, No. 5 (1958), 393-410.

[71] J.K. Knowles and E. Sternberg, On a class of conservation laws in linearized, and finite elasticity, Arch. Rational Mech. Anal. 44 (1972), 187.

[72] B. Vujanovič, Int. J. Non-linear Mech. 13,(1978), 185-197.

SHAPE OPTIMIZATION

OF

ELASTIC BARS IN TORSION

Jean W. Hou, Edward J. Haug, and Robert L. Benedict

Center for Computer Aided Design
College of Engineering
The University of Iowa
Iowa City, Iowa 52243

ABSTRACT

The problem of shape optimal design for multiply-connected elastic bars in torsion is formulated and solved numerically. A variational formulation for the equation is presented in a Sobolёv space setting and the material derivative idea of Continuum Mechanics is used for the shape design sensitivity analysis. The finite element method is used for a numerical solution of the variational state equation and is integrated into an iterative optimization algorithm. Numerical results are presented for both simply- and doubly-connected bars, with prescribed bounds on admissible location of both inner and outer boundaries.

ACKNOWLEDGEMENT

This research was supported by NSF Grant No. CEE 80-05677.

INTRODUCTION

The use of material derivative in the so called "speed method" was introduced by J. Cea, J.P. Zolesio, and B. Rousselet in a series of papers. For the details see [9], [7] and [8]. Here we apply it directly to the optimization of bars in torsion, and use the finite element technique to obtain specific numerical results for multiply-connected cross sections.

Consider the torsion problem for an elastic bar shown in Figure 1. The material of the bar is homogenous and isotropic and the cross section may have a void, thus resulting in a multiply-connected domain Ω. Torsional stiffness of the bar is defined by the following boundary-value problem for the stress function z (See reference [1]):

$$\left.\begin{aligned} \Delta z &= -2, && \text{in } \Omega && (1)\\ z &= 0, && \text{on } \Gamma_o && (2)\\ z &= q, && \text{on } \Gamma_i && (3) \end{aligned}\right\}$$

$$\int_{\Gamma_i} \frac{\partial z}{\partial n}\, dS = -2A_i \,, \qquad (4)$$

where Γ_o is the outer boundary of Ω and Γ_i is the inner boundary, enclosing the domain Ω_i. Here, the constant q is to be determined as part of the solution to the problem and A_i is the area of Ω_i. As shown in [1] the torsional rigidity is then given by

$$K = 2\iint_{\Omega} z \, d\Omega + 2qA_i = - \iint_{\Omega} x \cdot \nabla z \, d\Omega \quad , \tag{5}$$

where x is a position vector in Ω.

Polya and Weinstein [2], have proved the following assertion: "Of all doubly-connected cross sections with given areas of Ω and Ω_i, the ring bounded by two concentric circles has the maximum torsional rigidity."

Banichuk [3] and Kurshin and Onoprienko [4] have also investigated optimal shape of a bar with doubly-connected cross section. They hold the inner boundary Γ_i fixed and seek a shape for the outer boundary that maximizes torsional rigidity. The area of Ω is given. In addition to Equations 1-4, they obtain the following optimality condition for Γ_o:

$$\frac{\partial z}{\partial n} = C, \quad \text{on } \Gamma_o \quad . \tag{6}$$

Taking account of this excess condition, the boundary Γ_o is then determined so that the constant C matches the isoparametric constraint on area of Ω; i.e., the problem is treated as an inverse boundary-value problem. Banichuk uses a perturbation technique to obtain approximate solutions of this problem. He is able also to deduce some properties of an optimum contour. For example, wall thickness of the bar of optimum shape decreases as one moves along the inner boundary Γ_i in a direction of increasing curvature.

By restricting the cross section to be symmetric with respect to the coordinate axes, Kurshin and Onoprienko apply complex variable theory and solve Equations 1-6, with an isoparametric condition on the area of Ω. A system of nonlinear equations is obtained to determine the unknown coefficients of a complex function that describes the unknown boundary Γ_o. This system is solved by the Newton-Raphson method. Some numerical results are presented.

Quite recently, Dems [5] used a boundary perturbation analysis for a bar with doubly-connected cross section to maximize torsional rigidity, with the inner boundary held fixed. The optimality criteria obtained is the same as Equation 6. The shape optimization problem is formulated by defining shape of the boundary with a set of parameter-dependent, piecewise linear functions. The reduced problem is solved by means of the finite element method and an iterative algorithm based on the optimality condition. Several numerical examples are included.

The discussion thus far has focused on doubly-connected bars. If the cross section of the bar is simply-connected, Γ_i is a point ($A_i = 0$), and the value of q is immaterial. Thus, the boundary-value problem for the stress function z reduces to Equations 1 and 2 on the simply-connected domain Ω and the torsional rigidity is given by Equation 5. It is interesting that Equation 6 remains valid as an optimality criterion for the shape of Ω to maximize K with a given area of Ω [3].

Difficulties in solving the torsion problem for a bar with a doubly-connected cross section are associated with the boundary conditions of Equations 3 and 4. Usually, q in Equation 3 is determined from Equation 4. However, once an admissible function space and variational formulation can be defined, it is seen that Equation 4 becomes a defining equation for a natural boundary condition. Therefore, the Finite Element technique can be employed to solve the problem numerically. In section 2, such a variational formulation and admissible function space are defined and the equivalence between the variational formulation and the Equations 1-4. We also prove the existence and uniqueness of the solution.

In section 3, the material derivative concept is employed to obtain the directional derivative of torsional rigidity with respect to the shape of the domain by allowing both Γ_i and Γ_o to vary. Optimality criteria for the simply- and doubly- connected domains are obtained.

An iterative numerical method for optimizing shape of simply- and doubly-connected shaft cross sections is outlined in Section 4. Numerical calculations are carried out using the finite element method for analysis of the designs and a nonlinear programming method for optimization. Examples of both simply- and multiply-connected bars, with constraints on admissible location of the boundaries Γ_i and Γ_o, are presented in Section 5.

2. VARIATIONAL FORMULATION OF BOUNDARY-VALUE PROBLEMS

Suppose Ω is a doubly-connected open set in $\mathbb{R}^2$, bounded by regular boundaries Γ_i and Γ_o. The outer normals of the boundary curves are represented by n. The following bilinear and linear forms play a key role in the variational formulation of the problem:

$$a(z, v) = \int_\Omega \nabla z \cdot \nabla v \, d\Omega \tag{7}$$

$$(x, \nabla v) = \int_\Omega x \cdot \nabla v \, d\Omega \tag{8}$$

where $x \in \Omega \subset R^2$ is a position vector, ∇ denotes the gradient operator, and $x \cdot u = x_1u_1 + x_2u_2$. The variational equation for the torsion problem is given by:

$$a(z, v) + (x, \nabla v) = 0, \qquad \text{for all } v \in V, \tag{9}$$

where

$$V = \{v \in H^1(\Omega) \mid v = 0 \text{ on } \Gamma_o \text{ and } v = \beta \text{ on } \Gamma_i, \text{ for some } \beta \in \mathbb{R}^1\},$$

and where $H^1(\Omega)$ is the Sobolev space of order one [6].

One may define a formal operator Λ as, $\Lambda w(x) = -\Delta w(x)$, where $x \in \Omega$ and Δ is the Laplace operator. The domain of this formal operator is defined as

$$H^1(\Omega, \Lambda) = \{w \in H^1(\Omega) | \Lambda w \in L^2(\Omega)\} \tag{11}$$

Moreover, one may define the function space

$$V(\Lambda) = \{w \in H^1(\Omega, \Lambda) | w = 0 \text{ on } \Gamma_o \text{ and } w = \beta \text{ on } \Gamma_i, \text{ for some } \beta \in R^1\} \tag{12}$$

In the definitions of spaces in Equations 10-12, a function w evaluated on a boundary Γ is interpreted as a trace γw defined in $H^{\frac{1}{2}}(\Gamma)$ (See [6]). One may proceed to prove the following proposition:

Proposition. The following problems are equivalent:
Problem (a); Find $z \in H^1(\Omega, \Lambda)$ and $q \in R^1$ to satisfy

$$\Lambda z = 2, \quad \text{in } \Omega \tag{13}$$

$$z = 0, \quad \text{on } \Gamma_o \tag{14}$$

$$z = q, \quad \text{on } \Gamma_i \tag{15}$$

$$\int_{\Gamma_i} \frac{\partial z}{\partial n} dS + \int_{\Gamma_i} x \cdot \mathbf{n} \, dS = 0 \quad , \tag{16}$$

where $\int_{\Gamma_i} x \cdot \mathbf{n} \, dS$ is equal to twice of area enclosed in Γ_i.

Problem (b); Find $z \in V$ such that

$$a(z, v) + (x, \nabla v) = 0, \text{ for all } v \in V . \tag{17}$$

Proof:

(a → b). Suppose $z \in H^1(\Omega, \Lambda)$ is a solution of problem (a). Then, $z \in V(\Lambda)$. Since $z \in V(\Lambda) \subset H^1(\Omega, \Lambda)$ and $v \in V \subset H^1(\Omega)$, Green's formula for $z \in V(\Lambda)$ and any $v \in V$, is (as given in [6])

$$a(z, v) = (\Lambda z, v) + \langle \delta z, \gamma v \rangle$$

$$= (2, v) + \int_{\Gamma_i \; \Gamma_o} \frac{\partial z}{\partial n} v \, dS$$

$$= (2, v) + \beta \int_{\Gamma_i} \frac{\partial z}{\partial n} dS \tag{18}$$

where $\delta z \in H^{-1/2}(\Gamma)$ is an extension of $\frac{\partial z}{\partial n}$. The second and third equalities in Equation 18 are deduced from the facts that $\Lambda z = 2$ in Ω, $v = 0$ on Γ_o, and $v = \beta$ on Γ_i, for some constant β. Further, $\operatorname{div} x = \nabla \cdot x \equiv 2 \in L^2(\Omega)$. Hence, $x \in H^1(\Omega, \operatorname{div}) = \{u \in L^{2,2}(\Omega) \mid \operatorname{div} u \in L^2(\Omega)\}$ and

$$(x, \nabla v) = -(2, v) + \int_{\Gamma_i \cup \Gamma_o} x \cdot \mathbf{n}\, v\, dS$$

$$= -(2, v) + \beta \int_{\Gamma_i} x \cdot \mathbf{n}\, dS \tag{19}$$

Adding Equations 18 and 19 and using Equation 16, it follows that

$$a(z, v) + (x, \nabla v) = \beta\left(\int_{\Gamma_i} x \cdot \mathbf{n}\, dS + \int_{\Gamma_i} \frac{\partial z}{\partial n}\, dS\right) = 0 \tag{20}$$

for all $v \in V$. Thus, z is a solution of Problem (b).

(b $\rightarrow$ a). Suppose $z \in V$ is a solution of Problem (b). Since $z \in V$, the boundary conditions in Equations 14 and 15 of Problem (a) are satisfied. One may first consider only those $v \in V$ such that $v = 0$ on Γ_i; i.e., $v \in H_0^1(\Omega) \subset V \subset H^1(\Omega)$. Recall that $z \in V \subset H^1(\Omega)$ and $x \in H^1(\Omega, \operatorname{div})$. For this class of v, Green's formula (see [6]) is

$$a(z, v) = (\Lambda z, v) \quad , \tag{21}$$

where $\Lambda z \in H^{-1}(\Omega)$ and

$$(x, \nabla v) = (-2, v) \quad . \tag{22}$$

Adding Equations 21 and 22, it follows that

$$a(z, v) + (x, \nabla v) = (\Lambda z - 2, v)$$

for all $v \in H_0^1(\Omega)$. It is given that the left side of this equality vanishes, so $(\Lambda z-2, v) = 0$ for all $v \in H_0^1(\Omega)$. This implies that $\Lambda z - 2 = 0$ in $H^{-1}(\Omega)$. But $\Lambda z = 2 \in L^2$, so it follows that $z \in H^1(\Omega, \Lambda)$. Since $z \in H^1(\Omega, \Lambda)$, Green's formula is valid for all $v \in V \subset H^1(\Omega)$, giving

$$a(z, v) = (\Lambda z, v) + \int_{\Gamma_i \cup \Gamma_o} \frac{\partial z}{\partial n} v\, dS$$

$$= (\Lambda z, v) + \beta \int_{\Gamma_i} \frac{\partial z}{\partial n}\, dS \quad .$$

and

$$(x, \nabla v) = (-2, v) + \int_{\Gamma_i \cup \Gamma_o} x \cdot n \, dS$$

$$= (-2, v) + \beta \int_{\Gamma_i} x \cdot n \, dS$$

Adding, it follows that

$$a(z, v) + (x, \nabla v) = (\Lambda z-2, v) + \beta \left[\int_{\Gamma_i} (\frac{\partial z}{\partial n} + x \cdot n) dS\right], \text{ for all } v \in V .$$

The left side of this equation vanishes and it was shown that $\Lambda z = 2$, so

$$\beta \int_{\Gamma_i} (\frac{\partial z}{\partial n} + x \cdot n) dS = 0$$

for all $v \in V$, equivalently for any $\beta \in R^1$. Thus,

$$\int_{\Gamma_i} (\frac{\partial z}{\partial n} + x \cdot n) dS = 0$$

and the last condition (Equation 16) of Problem (a) is satisfied. Q.E.D.

One may define $\hat{\Omega} = \Omega \cup \Omega_i \cup \Gamma_i$ and extend the function $z \in V(\Omega)$ to z in $\hat{\Omega}$, with the definition

$$z = \begin{cases} z, & \text{in } \Omega \\ q \in R^1, & \text{in } \Omega_i \cup \Gamma_i \end{cases} .$$

An example of such a function is shown in Figure 2. This extended function belongs to $H_0^1(\hat{\Omega})$. For all $v \in H_0^1(\hat{\Omega})$, Poincare's inequality implies that

$$\int_{\hat{\Omega}} (\nabla v \cdot \nabla v) d\Omega > \alpha \int_{\hat{\Omega}} v^2 \, d\Omega \quad ,$$

where $\alpha > 0$. Adding $\hat{a}(v, v) \equiv \int_{\hat{\Omega}} \nabla v \cdot \nabla v \, d\Omega$ to both sides of the above inequality

and dividing by two, one has

$$\hat{a}(v, v) > \frac{\alpha}{2} \int_{\hat{\Omega}} v^2 \, d\Omega + \frac{1}{2} \int_{\hat{\Omega}} (\nabla v \cdot \nabla v) d\Omega$$

$$> \min\left(\frac{\alpha}{2}, \frac{1}{2}\right) ||v||^2_{H_0^1(\hat{\Omega})} \quad . \tag{23}$$

It is evident that $c = \min(\alpha/2, 1/2)$ is greater than zero for $\alpha > 0$. Because z is constant in Ω_i, $\nabla z = 0$ in Ω_i. Therefore,

$$\hat{a}(z, z) = \int_{\hat{\Omega}} \nabla z \cdot \nabla z \, d\Omega$$

$$= \int_{\Omega} \nabla z \cdot \nabla z \, d\Omega = a(z, z) \quad .$$

Furthermore,

$$||z||^2_{H_0^1(\hat{\Omega})} = ||z||^2_{V(\Omega)} + q^2(\text{mes } \Omega_i)$$

$$> ||z||^2_{V(\Omega)} \quad .$$

Substituting these results into Equation 23, one finally has $c > 0$ and

$$a(z, z) > c \, ||z||^2_V, \quad \text{for all } z \in V \quad . \tag{24}$$

Having proved V-ellipticity of a(z, z) (Equation 24), the Lax-Milgram Theorem (as used in [6]) ensures existence and uniqueness of a solution of the Problem (b). The proposition proved above implies that this solution is the unique solution of Problem (a).

3. SHAPE DESIGN SENSITIVITY ANALYSIS

Since the domain Ω is to be varied, it is convenient to treat it as a continuum and utilize the idea of material derivative, as introduced in continuum mechanics, to find the domain variation of the functionals concerned. One method of defining a variation in the domain Ω is to let $V(X)$, $X \in \Omega$, be a vector field that may be thought of as a "design velocity". A one parameter family perturbed domain may then be defined by the mapping

$$x = X + tV(X), \quad X \in \Omega, \ t \in R^1 \tag{25}$$

One may denote the deformed domain as $\Omega(t)$, with $x \in \Omega(t)$.

If z is the solution of Equation 17, which depends on the shape of $\Omega(t)$, then z depends on t both through the position $x = X + tV(X)$ and explicitly; i.e., $z = z(x, t)$. Under certain regularity hypothesis on Ω and the vector field $V(X)$ [7,8], one can define

$$\dot{z}(X) \equiv \lim_{t \to 0} \left[\frac{z(X+tV) - z(X)}{t}\right]$$

$$= z'(X) + \nabla z(X) \cdot \Lambda(X) \tag{26}$$

where $\dot{z}$ is the material derivative and z' is the partial derivative, defined as

$$z'(X) \equiv \lim_{t \to 0} \left[\frac{z(X, t) - z(X, 0)}{t}\right] \tag{27}$$

If $z \in H^1(\Omega)$, with smoothness assumptions on the domain and velocity field $V(X)$ [7], then $z' \in H^1(\Omega)$ [9], and $\dot{z} \in H^1(\Omega)$ [7, 8]. Thus, $(\nabla z \cdot V) \in H^1(\Omega)$. It is shown in References 7-9 that the following properties of the material derivative, which are well known in continuum mechanics, are valid in the Sobolev space setting:

$$(\nabla z)' = \nabla(z') \tag{28}$$

and for an integral functional

$$\dot{\Phi} = \iint_{\Omega} F(z, x)\, d\Omega \tag{29}$$

the material derivative is

$$\dot{\Phi} = \iint_{\Omega} \frac{\partial F}{\partial z} z'\, d\Omega + \int_{\Gamma_i \cup \Gamma_o} F\, V_n\, dS \tag{30}$$

where $V_n = V \cdot n$ is the normal component of V on the boundary of Ω.

More fundamental is the question of existence of the material and partial derivatives $\dot{z}$ and z' of the solution z of the variational equation of Equation 17. Under hypotheses of strong ellipticity of the energy bilinear form $a(\cdot, \cdot)$, proved in Equation 24, it is shown in References 8-10, that z is differentiable with respect to shape. With this knowledge and the material derivative formulas of Equations 26, 28, and 30, one can now study the torsional shape optimal design problem.

The first order domain variation of torsional rigidity is, from Equation 5 and 30,

$$\dot{K} = -(x, \nabla z') - \int_{\Gamma_i \cup \Gamma_o} x \cdot \nabla z \, V_n \, dS \tag{31}$$

Selecting v = z in Equation 17, one has,

$$a(z, z) + (x, \nabla z) = 0 \ . \tag{32}$$

Taking the material derivative of both sides of this equation gives

$$2a(z, z') + (x, \nabla z')$$

$$= - \int_{\Gamma_i \cup \Gamma_o} \nabla z \cdot \nabla z \, V_n \, dS - \int_{\Gamma_i \cup \Gamma_o} x \cdot \nabla z \, V_n \, dS \ . \tag{33}$$

Since $z' \in H^1(\Omega)$ and $\nabla z = -2$ in Ω, Green's formula yields

$$a(z, z') + (x, \nabla z') = (2, x') + \int_{\Gamma_i \ \Gamma_o} \frac{\partial z}{\partial n} z' \, dS$$

$$+ (-2, z') + \int_{\Gamma_i \cup \Gamma_o} (x \cdot n \, z') dS$$

$$= \int_{\Gamma_i \cup \Gamma_o} \frac{\partial z}{\partial n} z' \, dS + \int_{\Gamma_i \cup \Gamma_o} (x \cdot n \, z') dS \ .$$

Substituting this result into Equation 33, it follows that

$$2\int_{\Gamma_i \cup \Gamma_o} \frac{\partial z}{\partial n} z' \, dS + 2\int_{\Gamma_i \cup \Gamma_o} x \cdot z' \, dS - (z, \nabla z')$$

$$+ \int_{\Gamma_i \cup \Gamma_o} \nabla z \cdot \nabla z \, V_n \, dS + \int_{\Gamma_i \cup \Gamma_o} x \cdot \nabla z \, V_n \, dS = 0 \ . \tag{34}$$

On boundaries Γ_i and Γ_o, z is a constant, so $\frac{\partial z}{\partial s} = 0$. Furthermore, on Γ_o $\dot{z} = z' + \nabla z \cdot V = 0$, because z = 0 on Γ_o. However, on Γ_i $z = q$, $\dot{z} = z' + \nabla z \cdot V = \dot{q}$. It thus follows that

$$z' = \begin{cases} -\frac{\partial z}{\partial n} V_n, & \text{on } \Gamma_o \\ q - \frac{\partial z}{\partial n} V_n, & \text{on } \Gamma_i \end{cases}$$

and

$$x = \nabla z = x \cdot \mathbf{n} \frac{\partial z}{\partial n}, \text{ on } \Gamma_i \cup \Gamma_o \quad .$$

Therefore, Equation 31 becomes

$$\dot{K} = -(x, \nabla z') - \int_{\Gamma_i \cup \Gamma_o} x \cdot n \frac{\partial z}{\partial n} V_n \, dS \, , \tag{35}$$

and Equation 34 may be written as

$$-(x, \nabla z') - \int_{\Gamma_i \cup \Gamma_o} (\frac{\partial z}{\partial n})^2 V_n \, dS - \int_{\Gamma_i \cup \Gamma_o} x \cdot \mathbf{n} \frac{\partial z}{\partial n} V_n \, dS$$

$$+ 2\dot{q} \{ \int_{\Gamma_i} \frac{\partial z}{\partial n} dS + \int_{\Gamma_i} x \cdot n \, dS \} = 0 \quad . \tag{36}$$

Substituting from Equation 36 into Equation 35 and considering Equation 16, one has the desired result

$$\dot{K} = \int_{\Gamma_i \cup \Gamma_o} (\frac{\partial z}{\partial n}) V_n \, dS \quad . \tag{37}$$

Note that any monotone outward movement of the boundary; i.e., $V_n > 0$, yields an increase in K, which is to be expected. It is easy to repeat the arguments for a simply-connected domain Ω to see that Equation 37 is valid with Γ_i suppressed.

If the cross-sectional area A of the bar and the area A_i of the hole are given, isoparametric constraints on the shape of Ω are

$$\Phi = \int_{\Omega} d\Omega - A = 0 \tag{38}$$

$$\Phi_i = \int_{\Omega_i} d\Omega - A = 0 \tag{39}$$

Taking the derivative of both sides of these equations gives

$$\dot{\Phi} = \int_{\Gamma_i \cup \Gamma_o} V_n \, dS = 0 \qquad (40)$$

$$\dot{\Phi}_i = \int_{\Gamma_i} V_n \, dS = 0 \qquad (41)$$

The necessary condition for maximal torsional rigidity (equivalently the minimum of negative torsional rigidity), with shape variations consistent with Equations 38 and 39, is thus

$$-\int_{\Gamma_i \cup \Gamma_o} \left(\frac{\partial z}{\partial n}\right)^2 V_n \, dS + \lambda \int_{\Gamma_i \cup \Gamma_o} V_n \, dS + \lambda_i \int_{\Gamma_i} V_n \, dS = 0 \qquad (42)$$

for arbitrary V_n, where λ and λ_i are Lagrange multipliers corresponding to constraints of Equations 38 and 39. Under the assumption that V_n is smooth and arbitrary, provided no intersection of Γ_i and Γ_o occurs, one has the following necessary conditions of optimality:

$$-\left(\frac{\partial z}{\partial n}\right)^2 + \lambda + \lambda_i = 0, \quad \text{on } \Gamma_i \qquad (43)$$

$$\left(\frac{\partial z}{\partial n}\right)^2 + \lambda = 0, \quad \text{on } \Gamma_o \qquad (44)$$

It is clear that concentric circles for Γ_i and Γ_o satisfy these necessary conditions. This special case is proved in Reference 2.

If Γ_i is fixed, the necessary condition is only

$$-\left(\frac{\partial z}{\partial n}\right)^2 + \lambda = 0, \quad \text{on } \Gamma_o , \qquad (45)$$

which agrees with the results of Banichuk [3] and Dems [5].

Extensions of the preceding optimality conditions can be easily obtained using abstract optimization theory, in conjunction with the design sensitivity analysis results of this section. For example, if the inner boundary Γ_i is fixed and the outer boundary Γ_o is constrained to lie within some specified curve $\tilde{\Gamma}$, then at points on $\Gamma_o \cap \tilde{\Gamma}$ the only feasible variation of the domain is $V_n \leq 0$. Thus, one can prove existence of a multiplier function $\mu(X) > 0$, $X \in \Gamma_o \cap \tilde{\Gamma}$, such that Equation 42 on Γ_o becomes identified with the vanishing of the following boundary integral

$$\int_{\Gamma_o} [-(\frac{\partial z}{\partial n})^2 + \lambda + \mu(X)] V_n \, dS = 0$$

for arbitrary V_n on Γ_n. Thus, it is necessary that

$$\begin{aligned} -(\frac{\partial z}{\partial n})^2 + \lambda + \mu(X) &= 0, && \text{on } \Gamma_o \\ \mu(X) &> 0, && \text{on } \Gamma_o \cap \tilde{\Gamma} \\ \mu(X) &= 0, && \text{on } \Gamma_o / (\Gamma_o \cap \tilde{\Gamma}) \end{aligned} \tag{46}$$

While it is interesting to derive optimality conditions that must hold on the optimum boundary, such as Equations 43-46, it is difficlut to use these conditions to construct optimum shapes. One may view the necessary conditions as part of an inverse boundary-value problem; i.e., find the boundary of Ω so that the solution of a differential equation on Ω satisfies given boundary conditions and optimality criteria, such as Equations 43-46, on the boundary Γ. The latter, excess boundary conditions may be interpreted as determining the optimum location of the boundary. Banichuk approached a special case of the problem in this fashion in Reference 3, using a perturbation technique. Such methods are, however, very complicated and require a great deal of ad-hoc work for each problem treated.

A direct iterative optimization method is presented in the next section, based on the design sensitivity results obtained in this section, parameterization of the unknown boundary, and nonlinear programming methods.

4. ITERATIVE NUMERICAL SHAPE OPTIMAL DESIGN

A typical shape optimal design problem is to choose a domain Ω to minimize a cost functional of the form

$$\Phi_0 = \iint_{\Omega} G^0(z) \, d\Omega \quad , \tag{47}$$

subject to functional constraints

$$\Phi_i = \iint_{\Omega} G^i(z) \, d\Omega \begin{cases} = 0, & i = 1, \ldots, k' \\ < 0, & i = k' + 1, \ldots, k \end{cases} , \tag{48}$$

where the state z is the solution of a variational equation of the form of Equation 17. It is further required that the boundary Γ lie between Γ^+ and Γ^-, as shown in Figure 3. The latter pointwise constraints are written in the form

$$\left.\begin{aligned} d^n(\Gamma, \Gamma^+) &> 0 \\ d^n(\Gamma^-, \Gamma) &> 0 \end{aligned}\right\} \tag{49}$$

where $d^n(.,.)$ is the distance measured along the normal n to Γ, from the first to the second curve.

Using results of Equations 37, 40, and 41, each of the functionals of Equations 47 and 48 can be differentiated (linearized) to obtain

$$\delta\Phi = \int_\Gamma \Lambda^i V_n \, d\Gamma, \qquad i = 0, 1, \ldots, k \tag{50}$$

where the sensitivity coefficients Λ^i of V_n in Equations 37, 40, and 41 define variations in the cost functional and each active functional constraint.

Even though the linearized functional appearing in Equation 50 has been obtained using a variational formulation, a finite dimensional parameterization of the boundary can be introduced to reduce this linearized functional to parametric form. Presume that points on the boundary Γ are specified by a vector $r(\alpha;b)$ from the origin of the coordinate system to the boundary, as shown in Figure 4, where α is a parameter vector and b is a vector of design parameters $b = [b_1, \ldots, b_m]^T$. When the parameterization of Γ has been defined, the domain optimization problem reduces to selection of the finite dimensional vector b to minimize the cost function of Equation 47, subject to the constraints of Equations 48 and 49. The linearized form of this problem may be written in terms of variation δb by denoting

$$b = b^0 + t\delta b \,, \tag{51}$$

where b^0 is the design at a given iteration. The velocity field at the boundary is

$$V = \frac{d}{dt}\,(r(\alpha;b)) = \frac{\partial r}{\partial b}\,\delta b \,. \tag{52}$$

Taking the dot product of V with the unit outward normal to the curve Γ yields

$$V_n = n \cdot V = [n \cdot \frac{\partial r(\alpha;b)}{\partial b}]\,\delta b \,. \tag{53}$$

Here, the coefficient of δb can be calculated at each point on Γ and the result substituted into Equation 50 to obtain

$$\delta\Phi_i = [\int_\Gamma \Lambda^i \,(n \cdot \frac{\partial r}{\partial b}\,)\, dS]\delta b \equiv \ell^{i^T} \delta b \,. \tag{54}$$

More directly, the pointwise constraints on location of Γ in Equation 49 can be linearized as

$$-d^n(\Gamma(\alpha), \Gamma^+) < n(\alpha) \cdot \frac{\partial r}{\partial b}(\alpha)\, \delta b < d^n(\Gamma^-, \Gamma(\alpha)), \text{ on } \Gamma \; . \tag{55}$$

This constraint may be implemented over Γ in several ways, the simplest being to enforce it at a grid α_j of points.

Having defined a finite dimensional parameterization of the shape optimal design problem and obtained derivatives of the cost and constraint functions with respect to design parameters, one can now apply any well known nonlinear programming algorithm to iteratively optimize the shape. In each iteration, a finite element approximate solution of the boundary-value problem is constructed and used to evaluate design derivatives of torsional stiffness, using Equations 37 and 54. More directly, Equations 40, 41, and 54 are used to calculate derivatives of Φ and Φ_i in Equations 38 and 39. Finally, the derivatives appearing in Equation 55 are calculated directly.

Numerical results presented in the following section have been obtained by a recursive quadratic programming algorithm [11] that has been proved to be globally convergent [12]. With the design derivatives calculated, however, any gradient based, nonlinear programming algorithm can be used.

5. NUMERICAL EXAMPLES

Example 1

The first example presented deals with Polya and Weinstein's proof that concentric circles define the optimum shape, if no constraints are placed on boundary location. The amount of material is given as 65 units and the area of the hole is 20 units. Both conditions are treated as isoparametric constraints. As an initial design, two concentric circles are selected with radii 4.5 and 2.0 units, respectively. A regular polygon is used to approximate the boundary, as shown in Figure 5. The radial distances b_i between the i-th vertex and the origin are chosen as design variables.

For the coarse grid model (96 elements, 64 nodes, and 16 design variables in Figure 5(a)), six iterations, requiring 7.93 CPU seconds on a PRIME 750 minicomputer, were required for convergence to the optimum shape. It took 7 iterations and 51.57 CPU seconds for the finer grid model (384 elements, 224 nodes, and 64 design variables in Figure 5(b)) to achieve convergence. A comparison between the theoretical values and the final optimum results is given in Table 1.

Example 2

As a second example, the inner boundary is fixed as an ellipse with semi-radii a = 2.5 and b = 1.0 units. The amount of material is given as 45 units. The initial estimate for the outer boundary was taken as a circle with radius 4.5 units. Thirty-five iterations and 270.1 CPU seconds on a PRIME 750 minicomputer were equired to achieve convergence to the design shown in Figure 6. The torsional rigidity is 415.83 at the final solution, while the initial value is 604.74. These results support Banichuk's claim that wall thickness of the bar at the optimum shape decreases as one moves along the inner boundary in a direction of increasing curvature.

Example 3

As a final example, both the outer and inner boundaries are treated as design variables. In addition to the constraint on the amount of material available, the cross section of the bar is required to be in a 10 x 16 unit rectangular housing. Two finite element meshes are used for analysis. One has 384 elements, 224 nodes, and 64 design variables as in the preceding example. The second mesh has 960 elements, 528 nodes, and 96 design variables. The initial design is taken as two concentric circles of radii 5, and 2.5 units.

With given amounts of material of 85 and 110 units, numerical results are listed in Table 2. Optimum shapes, for different finite element meshes, are shown in Figures 7 and 8. Note that the corners of the housing are not filled for all examples, as one might expect. Although the values of optimum torsional rigidities are very close for the two finite element meshes, optimum shapes of the inner boundaries show significant differences. It is apparent that improved stress evaluation, which gives a better approximation of design sensitivity coefficients, has caused this deviation. It is also interesting to see that the bar with a hole has distributed the material more efficiently (has higher torsional rigidity) than the bar with a solid cross section. Calculated with a finite element model of 384 elements, 209 nodes, and 32 design variables, numerical results for the optimum design of a solid bar are listed in Table 3. The optimum shapes are shown in Figure 9.

6. CONCLUSIONS AND REMARKS

The numerical examples offered here illustrate the wide applicability of the iterative numerical schemes for shape optimal design. Note that the pattern of finite element mesh does not change during an iteration. In each iteration, new positions of boundary nodes are determined by the algorithm and the positions of interior nodes change accordingly.

The sensitivity functional, derived using the variational formulation of the state equation and material derivative, is a boundary integral that contains only the normal boundary movement (that is V_n) and the stress terms $(\frac{\partial z}{\partial n})$. Success in a numerical technique for shape optimal design depends on an accurate evaluation of these stress terms and on the representation of the boundary and its normal movement.

A more sophisticated choice of elements or of a finer mesh can be introduced in the finite element method to improve the numerical approximation of stress values. Instead of linear piecewise functions, some smoother or more restricted classes of functions may be used to describe the boundary shapes. From engineer's point of view this will undoubtedly broaden the utilization of the shape optimization techniques. A detailed discussion of some mathematical approaches to the choices of finite elements and to the grid optimization for the finite element formulation of structural problems is offered in the paper of A.R. Diaz, N. Kikuchi and J.E. Taylor [14] in this volume. Also see [12] and [8]. For a basic introduction to this topic, see reference [13].

Finally, we comment that the basic problem of pure torsion of an elastic multiply connected bar is an important problem in the theory of elasticity and does have a long history. Large body of literature concerning it goes back to the original papers of Saint Venant, Lord Kelvin (Sir William Thompson) and Prandtl. While the numerical aspects and the theoretical results of this paper dealt with a version of this classical problem or rather with the related problem of shape optimization for elastic multiply connected bars subjected to pure torsion, the uses of material derivative and the other concepts utilized here are quite general and are certainly not restricted to the specific problem stated in the title of this paper.

An elementary introduction to the theory of pure torsion for linear elastic isotropic bars may be found in the reference [1].

REFERENCES

1. Sokolnikoff, I.S., Mathematical Theory of Elasticity, McGraw-Hill, New York, (1956).

2. Polya, G., and Weinstein, A., "On the Torsional Rigidity of Multiply Connected Cross-Sections", Ann. of Math, 52, 154-163, (1950).

3. Banichuk, N.V., "Optimization of Elastic Bars in Torsion", Int. J. Solids Struct., 12, 275-286, (1976).

4. Kurshin, L. M., and Onoprienko, P.N., "Determination of the Shapes of Doubly-Connected Bar Sections of Maximum Torsional Stiffness", Prik. Math. Mech. (English translation Appl. Mathematics and Mechanics, PMM), 40, 1078-1084, (1976).

5. Dems, K., "Multiparameter Shape Optimization of Elastic Bars in Torsion", Int. J. Num. Meth. Engng., 15, 1517-1539, (1980).

6. Aubin, J.P., Applied Functional Analysis, Wiley-Interscience, New York, (1979).

7. Zolesio, J.P., "The Material Derivative (or speed) Method for Shape Optimization", Optimization of Distributed Parameter Structures, Vol. II, (ed, E.J. Haug and J. Cea) Sitjthoff-Noordhoff, Rockville, Md., 1089-1151, (1981).

8. Haug, E.J., Choi, K.K., and Komkov, V., Structural Design Sensitivity Analysis, Academic Press, New York, (1984).

9. Cea, J., "Problems of Shape Optimal Design", Optimization of Distributed Paramemter Structures, Vol. II. (ed. E.J. Haug and J. Cea) Sitjthoff-Noordhoff, Rockville, Md., 1005-1048, (1981).

10. Rousselet, B., and Haug, E.J., "Design Sensitivity Analysis of Shape Variation", Optimization of Distributed Parameter Structures, Vol. II, (ed. E.J. Haug and J. Cea) Sijthoff-Noordhoff, Rockville, Md., 1397-1442, (1981).

11. Choi, K.K., Haug, E.J., Hou, J.W., and Sohoni, V.N., "Pshenichny's Linearization Method for Mechanical System Optimization", Trans, ASME, J. Mech. Design, to appear (1984).

12. Pshenichny, B.N., and Danilin, Y.M., Numerical Methods in Extremal Problems, Mir, Moscow, (1978).

13. Zienkiewicz, O.C., The Finite Element Method in Engineering Science, (the second expanded edition), McGraw Hill, London, 1971.

14. Diaz, A.R., Kikuchi N. and Taylor J.E., Optimal Design Formulation for Finite Element Grid Adaptation, in this volume.

LIST OF FIGURES AND TABLES

Figure

1 Torsion of a Doubly-Connected Bar

2 Stress Function for a Doubly-Connected Bar

3 Pointwise Constraint on Boundary Γ

4 Parametric Definition of Γ

5 Finite Element Methods of Elastic Bar's Cross Section
(a) Coarse Grid Model
(b) Fine Grid Model

6 Final Optimum Shape

7 Final Optimum Shapes for a Torsion Bar with Coarse Mesh
(a) Given Amount of Material is 85 Units
(b) Given Amount of Material is 110 Units

8 Final Optimum Shapes for a Torsion Bar with Fine Mesh
(a) Given Amount of Material is 85 Units
(b) Given Amount of Material is 110 Units

9 Final Optimum Shapes for a Torsion Bar with Solid Cross Section
(a) Given Amount of Material is 85 Units
(b) Given Amount of Material is 110 Units

Table

1 Numerical Results for Optimum Shapes

2 Numerical Results for a Torsion Bar with Hollow Cross-Section

3 Numerical Results for a Torsion Bar with Solid Cross-Section

Table 1. Numerical Results for Optimum Shapes

	Theoretical Values	Optimum Values	
		Coarse Grid	Fine Grid
Torsional Rigidity	1086.44	1067.4	1081.51
Radius of Outer Boundary	5.2016	$5.2625 < b_1 < 5.2716$	$5.2180 < b_i < 5.2186$
Radius of Inner Boundary	2.523	$2.523 < b_1 < 2.5570$	$2.5312 < b_i < 2.5313$

Table 2. Numerical Results for a Torsion Bar with Hollow Cross-Section

Given Material	No. of Iterations for Convergence		CPU Seconds on PRIME 750		Optimum Torsional Rigidity		Constraint on Area	
	Coarse Grid	Finer Grid	Coarse Grid	Finer Grid	Coarse Grid	Finer Grid	Coarse Grid	Finer Grid
85 units	704	410	1419.2	1602.7	2457.3	2433.4	83.79	82.9
110 units	755	410	1523.2	1602.7	2826.9	2820.8	107.8	106.7

Table 3. Numerical Results for a Torsion Bar with Solid Cross-Section

Given Material	No. of Iterations for Convergence	CPU Seconds on PRIME 750	Optimum Torsional Rigidity	Constraint on Area
85 units	47	247.4	1139.3	84.99
110 units	177	923.4	1785.2	108.1

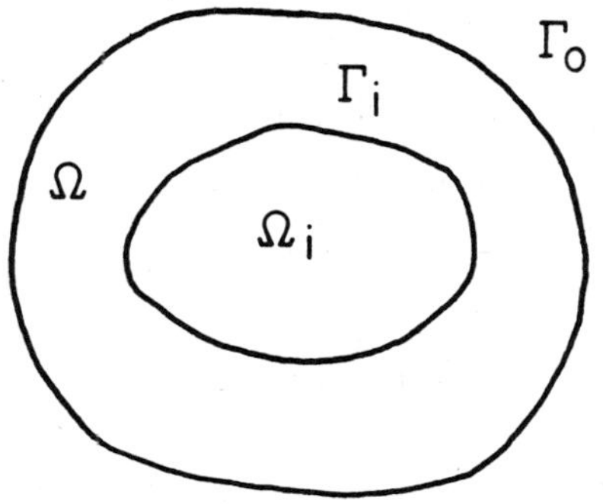

Figure 1. Torsion of a Doubly-Connected Bar

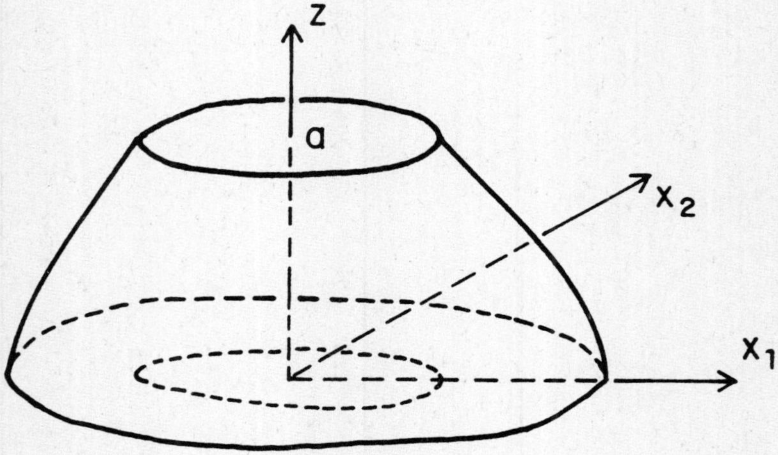

Figure 2. Stress Function for a Doubly-Connected Bar

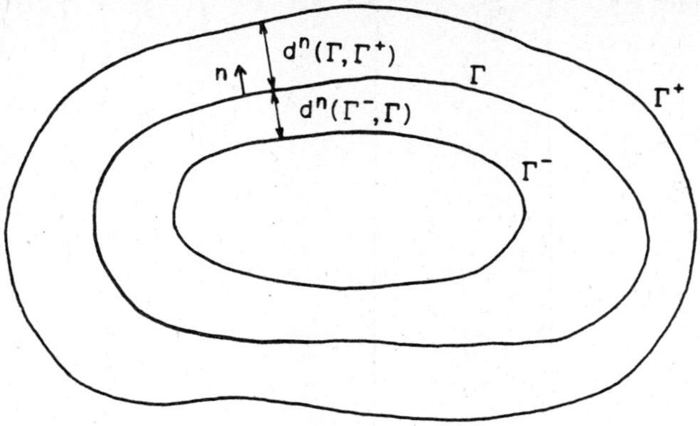

Figure 3. Pointwise Constraint on Boundary Γ

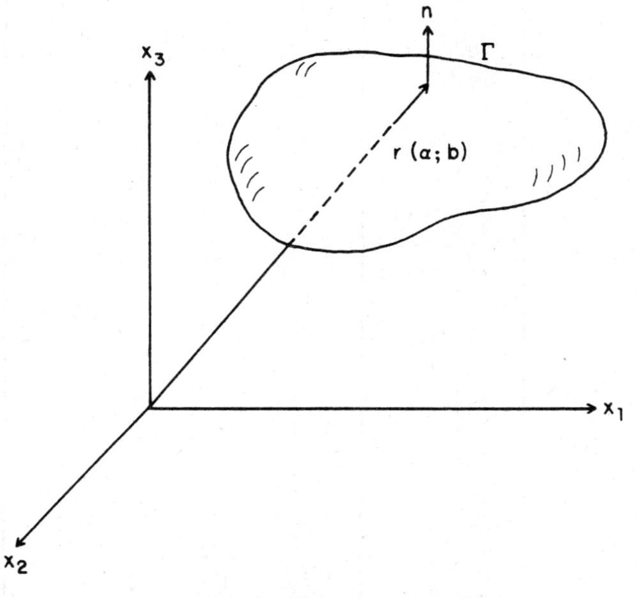

Figure 4. Parametric Definition of Γ

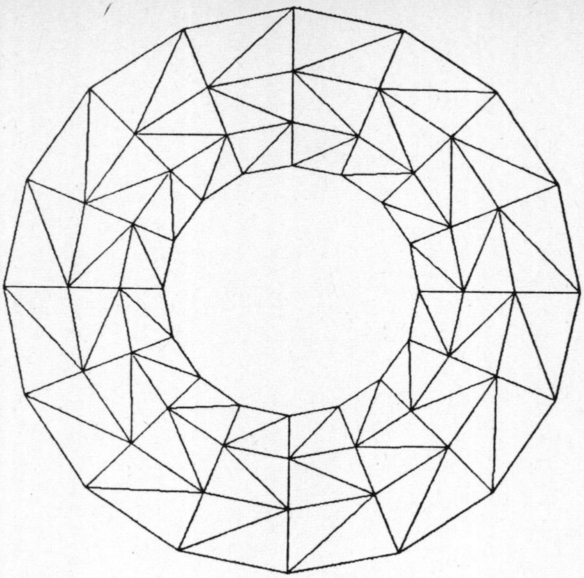

(a) Coarse Grid Model

(b) Fine Grid Model

Figure 5. Finite Element Models of Elastic Bar's Cross Section

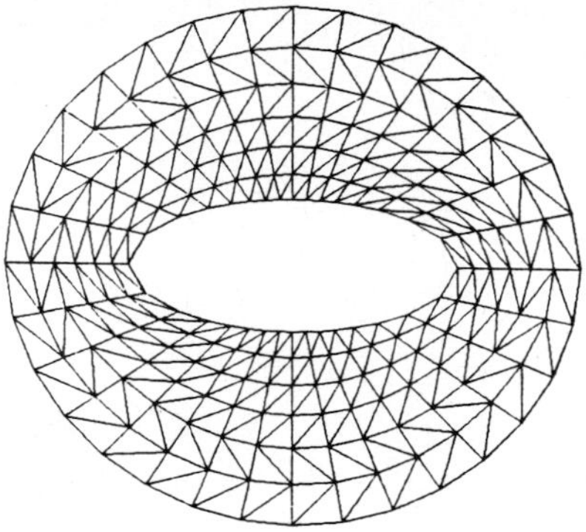

Figure 6. Final Optimum Shape

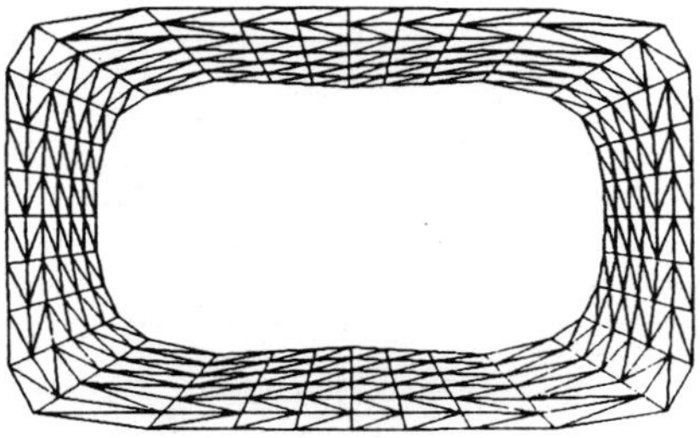

(a) Given amount of material is 85 units

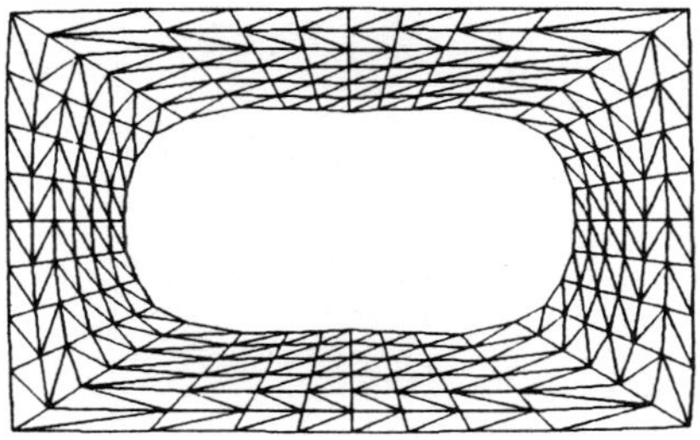

(b) Given amount of material is 110 units

Figure 7. Final Optimum Shapes for a Torsion Bar with Coarse Mesh

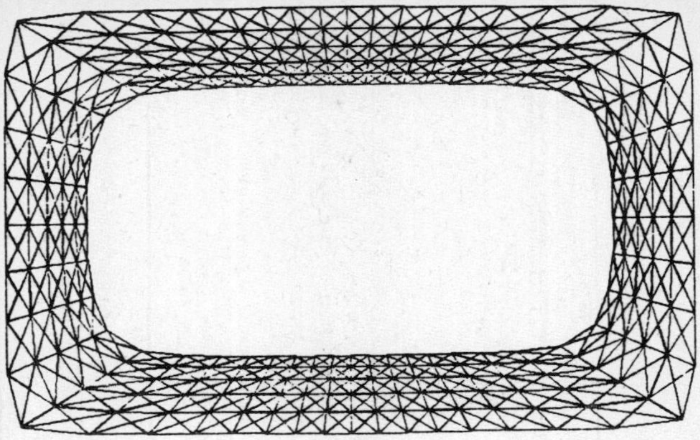

(a) Given amount of material is 85 units

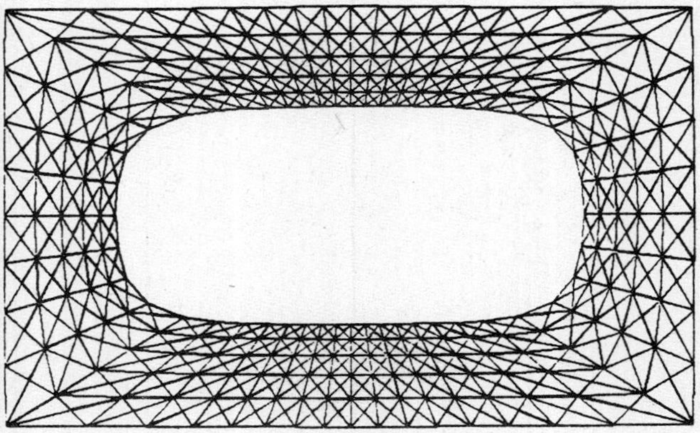

(b) Given amount of material is 110 units

Figure 8. Final Optimum Shapes for a Torsion Bar with Fine Mesh

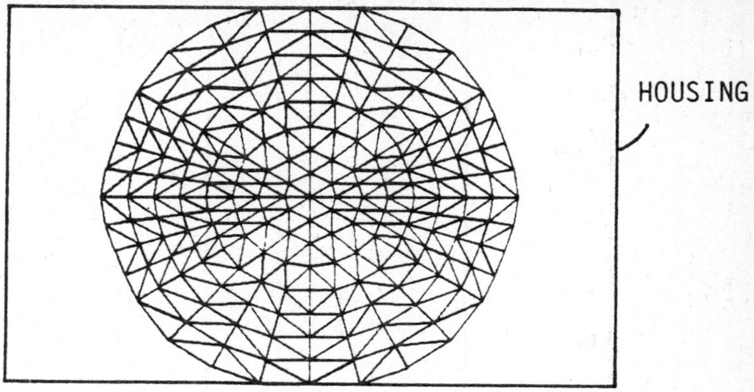

(a) Given amount of material is 85 units

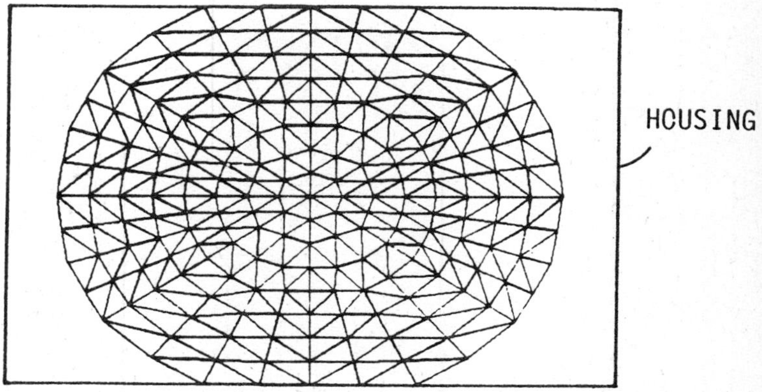

(b) Given amount of material is 110 units

Figure 9. Final Optimum Shapes for a Torsion Bar with Solid Cross-Section

OPTIMAL DESIGN FORMULATIONS FOR FINITE ELEMENT GRID ADAPTATION

Alejandro R. Diaz
Intevep/Ingenieria General
Apartado 76343, Caracas 1080A, Venezuela

Noboru Kikuchi
J. E. Taylor
College of Engineering, The University of Michigan
Ann Arbor, MI 48109-2140 U.S.A.

Grid adaptation procedures are used to obtain *improved quality* in finite element method computational results, via the introduction of appropriate *adjustments* in the *grid* or *mesh configuration*. Such procedures might be used to avoid difficulties otherwise encountered in situations where sharp variations occur in the field variable or in its derivatives; contact problems, or problems with stress concentrations in elasticity are examples. Recent experience with shape optimization problems suggests that the incorporation of means for grid adaptation is important in that application as well. In this paper we address issues related to the formulation and solution of grid adaptation problems, treated in the context of methods for structural design optimization.

There exist extensive works on the topics of mesh generation and mesh adaptation for finite difference modelling in computational fluid dynamics, e.g., Babuska, Chandry and Flaherty [1] and Ghia and Ghia [2]. By comparison, relatively little has been published on grid modification methods in conjunction with finite element methods. Developments on methods in general for the improvement of finite element computational results are summarized to a large extent in the works of Oliveira [3], Prager [4], Tang and Turcke [5], Masur [6], Melosh and Marcal [7], Turcke and McNeice [8], Shephard and Gallagher [9], Babuška and Rheinboldt [10-12], Babuška and Szabo [13], Sewell [14], Zienkiewicz, Gago and Kelly [15], Fellippa [16,17], Shephard [18], and Rossow and Katz [19]. These papers cover techniques for increasing the number of elements (h-methods), approaches to improve results by increasing the degree of polynomial in shape functions (p-method), as well as the grid modification methods. Techniques for systematic improvement of computational results by 'h' and 'p' methods were investigated in depth by Babuška and his co-workers, and by other investigators. On the other hand, mathematical justification of procedures for grid optimization has been lacking.

As already noted, grid adaptation is approached in this study as a problem in optimal design. Consideration is given to the view of how properly to characterize measures of the quality of computational results, to serve as a basis for optimization. A variational formulation is given for the optimal grid design problem, and the associated necessary conditions are derived. Based on these necessary conditions, an algorithm is given for the prediction of optimal relocation of interior nodes. The solution procedure is discussed and demonstrated for examples of analysis for the stress field in a membrane with notches, for the stress analysis of a nonhomogeneous structure, and for an example solution of the Navier-Stokes equation.

Design of Finite Element Grids

As indicated in the introduction, our purpose is to establish models for the problem of optimizing FEM grids within a fixed number of elements and a specified degree of polynomial shape function. The optimal design problem is expressed as (An extensive study of grid design based on a global measure of error is reported in [20]):

$$\min \left[\max_{e=1,2,\ldots,E} E_e \right] \tag{1}$$

where E_e represents a nonnegative measure of error in the eth element, E equals the total number of elements, and the minimization is with respect to locations of the nodes. Measure E_e is chosen according to the purpose or the practical requirements of the analysis. Of course the original domain covered by the grid is preserved. The 'max norm' represented in (1) reflects the quality of computational results in each portion (element) of the structure throughout its domain, i.e., the maximum of E_e bounds the chosen local measure of error.

Since E_e, e=1,2,...,E are nonnegative, a necessary conditions for the min/max problem (1) is obtained as

$$E_e = \text{constant}, \qquad e=1,2,\ldots,E \tag{2}$$

for the optimal location of the nodes of the finite element grid. If condition (2) is not met, then it is possible in general to reduce the maximum value of error by making an adjustment of the grid. Issues related to sufficiency for the solution of problem (1) are not addressed here. Also we note that condition (2) may not imply uniqueness for either the location of nodes or the shape of elements.

Next a review of errors in finite element approximations is given, to provide background for the definition of appropriate local measures E_e. In this discussion it is supposed that the original field problem is described by the abstract form

$$\underset{\sim}{u} \in V:\ a(\underset{\sim}{u},\underset{\sim}{v}) = f(\underset{\sim}{v}), \quad \text{for all } \underset{\sim}{v} \in V_o \tag{3}$$

where $\underset{\sim}{V}$ is a linear manifold of admissible displacements satisfying possibly non-homogeneous kinematic boundary conditions, $\underset{\sim}{V}_o$ is its homogeneous version, $a(\underset{\sim}{u},\underset{\sim}{v})$ is the (internal) virtual work by an arbitrary virtual displacement $\underset{\sim}{v}$ at the equilibrium position defined by the displacement $\underset{\sim}{u}$, and $f(\underset{\sim}{v})$ is the 'work' of applied body forces and boundary tractions associated with virtual displacement $\underset{\sim}{v}$. For linearly elastic bodies, $a(\cdot,\cdot)$ represents a bilinear form and $f(\underset{\sim}{v})$ is a linear functional in $\underset{\sim}{v}$.

A finite element approximation of (3) is given by

$$\underset{\sim}{u}_h \in \underset{\sim}{V}_h:\ a(\underset{\sim}{u}_h,\underset{\sim}{v}_h) = f(\underset{\sim}{v}_h), \quad \text{for all } \underset{\sim}{v}_h \in \underset{\sim}{V}_{oh} \tag{4}$$

where $\underset{\sim}{V}_h$ and $\underset{\sim}{V}_{oh}$ are finite element approximations of sets of nonhomogeneous and homogeneous admissible displacements. A component of a displacement $\underset{\sim}{v}_h \varepsilon \underset{\sim}{V}_h$ is a piecewise polynomial constructed by a linear combination of its values at nodes and shape functions defined in each element. Dimension of the space $\underset{\sim}{V}_h$ equals the total number of degrees of freedom, i.e., the total number of measures for nodal displacements. Since $\underset{\sim}{V}$ can be obtained as the closure of the union of $\underset{\sim}{V}_h$ for $0 < h < \infty$, where h is the representative size of finite elements, it is asymptotically true that a greater number of degrees of freedom implies a better representation of $\underset{\sim}{V}$. That is, if $h \to 0$ then $\underset{\sim}{V}_h \to V$.

The degree of approximation of $\underset{\sim}{v}$ thus depends on the total number of degrees of freedom in a finite element discretization Ω_h of the domain Ω, i.e., it depends on the total number of finite elements Ω_e, e=1,...,E, such that $\Omega_h \bigcup_{e=1}^{E} \Omega_e$. However, this does not imply any quantitative measure for the degree of approximation. That is, although it is possible to say that $\underset{\sim}{u}_h \to u$ as $h \to 0$, it is not known how close $\underset{\sim}{u}_h$ is to the true solution $\underset{\sim}{u}$ for a "given" h, i.e., for a given total number of degrees of freedom. To provide a quantitative measure, let us note the following inequality:

$$\sqrt{a(\underset{\sim}{u}-\underset{\sim}{u}_h,\ \underset{\sim}{u}-\underset{\sim}{u}_h)} \leq \sqrt{a(\underset{\sim}{u}-\underset{\sim}{v}_h,\underset{\sim}{u}-\underset{\sim}{v}_h)} \quad \text{for any } \underset{\sim}{v}_h \varepsilon \underset{\sim}{V}_h \,. \tag{5}$$

Indeed, putting $\underset{\sim}{v}_h$ into $\underset{\sim}{v}$ in (4) and subtracting (5) yields

$$a(\underset{\sim}{u}-\underset{\sim}{u}_h,\underset{\sim}{v}_h) = 0, \quad \text{for all } \underset{\sim}{v}_h \varepsilon \ \underset{\sim}{V}_{oh}. \tag{6}$$

On the other hand, we have, for any $\underset{\sim}{v}_h \varepsilon \underset{\sim}{V}_h$,

$$a(\underset{\sim}{u}-\underset{\sim}{u}_h,\underset{\sim}{u}-\underset{\sim}{u}_h) = a(\underset{\sim}{u}-\underset{\sim}{u}_h,\underset{\sim}{u}-\underset{\sim}{v}_h) + a(\underset{\sim}{u}-\underset{\sim}{u}_h,\underset{\sim}{v}_h-\underset{\sim}{u}_h).$$

Noting that $\underset{\sim}{v}_h-\underset{\sim}{u}_h \varepsilon \underset{\sim}{V}_{oh}$, (6) yields

$$a(\underset{\sim}{u}-\underset{\sim}{u}_h,\underset{\sim}{u}-\underset{\sim}{u}_h) = a(\underset{\sim}{u}-\underset{\sim}{u}_h,\underset{\sim}{u}-\underset{\sim}{v}_h). \tag{7}$$

Applying the Schwarz inequality under an additional assumption on the bilinear form $a(\cdot,\cdot)$ such that

$$a(\underset{\sim}{w},\underset{\sim}{w}) \geq 0, \quad \text{for all } \underset{\sim}{w} \varepsilon \underset{\sim}{V} \text{ and } a(\underset{\sim}{w},\underset{\sim}{w}) = 0 \text{ implies } \underset{\sim}{w} = \underset{\sim}{0} \tag{8}$$

we have

$$\sqrt{a(\underset{\sim}{u}-\underset{\sim}{u}_h,\underset{\sim}{u}-\underset{\sim}{u}_h)} \leq \sqrt{a(\underset{\sim}{u}-\underset{\sim}{u}_h,\underset{\sim}{u}-\underset{\sim}{u}_h)} \sqrt{a(\underset{\sim}{u}-\underset{\sim}{v}_h,\underset{\sim}{u}-\underset{\sim}{v}_h)}$$

Therefore, if an error of finite element approximations is defined by

$$e_h = \sqrt{a(\underset{\sim}{u}-\underset{\sim}{u}_h,\ \underset{\sim}{u}-\underset{\sim}{u}_h)},$$

we have the relation

$$e_h^2 \leq a(\underset{\sim}{u}-\underset{\sim}{v}_h,\ \underset{\sim}{u}-\underset{\sim}{v}_h), \quad \text{for all } \underset{\sim}{v}_h \varepsilon \underset{\sim}{V}_h. \tag{9}$$

Since $\underset{\sim}{v}_h$ in the right hand side of (9) is arbitrary, it may be identified with the *interpolation* of $\underset{\sim}{u}$ in the finite element approximation $\underset{\sim}{v}_h$ of the admissible set $\underset{\sim}{V}$ of displacements. This choice makes it possible to establish a quantitative measure of error in finite element approximations. Indeed, the error e_h due to finite element approximation is *a priori* bounded by the error of the interpolation according to inequality (5) or (9) by choosing $\underset{\sim}{v}_h$ as the interpolation of the solution $\underset{\sim}{u}$ of the original problem (3). Noting that the right hand side of (9) can be written as

$$a(\underset{\sim}{u}-\underset{\sim}{v}_h,\ \underset{\sim}{u}-\underset{\sim}{v}_h) = \sum_{e=1}^{E} a_e(\underset{\sim}{u}-\underset{\sim}{v}_h,\ \underset{\sim}{u}-\underset{\sim}{v}_h) \tag{10}$$

where $a_e(\underset{\sim}{u},\underset{\sim}{v})$ is the (internal) work of the finite element Ω_e, $e=1,\ldots,E$, one of the choices of error measures is

$$E_e = \sqrt{a_e(\underset{\sim}{u}-\underset{\sim}{v}_h,\ \underset{\sim}{u}-\underset{\sim}{v}_h)}. \tag{11}$$

This error measure represents the square root of twice the strain energy in a finite element Ω_e, associated with the difference between the true solution u and its finite element interpolation $\underset{\sim}{v}_h$.

Also, if the number E of elements is finite, then

$$e_h \le \sqrt{a(\underset{\sim}{u}-\underset{\sim}{v}_h,\ \underset{\sim}{u}_h-\underset{\sim}{v}_h)} \le \sqrt{E} \max_{e=1,2,\ldots,E} E_e \tag{12}$$

Equality holds in the second of these relations only if

$$E_e = \text{constant}, \qquad e=1,2,\ldots,E$$

But this is simply the necessary condition given in equation (2) for the grid design problem (1). Since for the optimal grid the $'\max_{e=1,2,\ldots,E} E_e'$ (for the E_e of (11)) is minimized, it follows that the interpolation error e_h^I defined by ($\underset{\sim}{v}_h$ is the interpolation of $\underset{\sim}{u}$)

$$e_h^I = \sqrt{a(\underset{\sim}{u}-\underset{\sim}{v}_h,\ \underset{\sim}{u}-\underset{\sim}{v}_h)} \tag{13}$$

is minimized for the optimal finite element grid. The implication of this result on the measure e_h of finite element error follows from the first inequality in (12).

Other Choices of Error Measure

One of the key ideas in the discussion of the previous section is that the error due to finite element approximation e_h is bounded by the interpolation error e_h^I. These respective errors are defined in terms of strain energy for the elastic structural system. We now postulate that the error of finite element approximation is bounded by interpolation error for other measures of error as well. For example, suppose that the finite element error is defined by

$$e_h = \{\int_\Omega \sigma_{ij}(\underset{\sim}{u}-\underset{\sim}{u}_h)\sigma_{ij}(\underset{\sim}{u}-\underset{\sim}{u}_h)d\Omega\}^{1/2} \tag{14}$$

where $\underset{\sim}{\sigma} = \sigma_{ij}\underset{\sim}{e}_i\underset{\sim}{e}_j$ is the stress tensor of a linearly elastic body, and the summation convention applies. We shall assume that there exists a positive constant C independent of the discretization of the domain such that

$$C\, e_h \le e_h^I \triangleq \{\int_\Omega \sigma_{ij}(\underset{\sim}{u}-\underset{\sim}{v}_h)\sigma_{ij}(\underset{\sim}{u}-\underset{\sim}{v}_h)d\Omega\}^{1/2} \tag{15}$$

Here $\underset{\sim}{v}_h$ is the interpolation of the solution $\underset{\sim}{u}$ of (4). With the assumption of linear elasticity,

$$\varepsilon_{ij} = C_{ijk\ell}\sigma_{k\ell} \tag{16}$$

where $\underset{\sim}{C}$ is the compliance matrix. Then the bilinear form $a(\cdot,\cdot)$ is given by

$$\begin{aligned} a(u,v) &= \int_\Omega \sigma_{ij}(\underset{\sim}{u})\varepsilon_{ij}(\underset{\sim}{v})d\Omega \\ &= \int_\Omega \sigma_{ij}(\underset{\sim}{u})C_{ijk\ell}\sigma_{k\ell}(\underset{\sim}{v})d\Omega \end{aligned} \tag{17}$$

Thus if we assume the boundedness of the compliance matrix $\underset{\sim}{C}$ such that

$$m\sigma_{ij}\sigma_{ij} \leq C_{ijk\ell}\sigma_{k\ell}\sigma_{ij} \leq M\sigma_{ij}\sigma_{ij} \tag{18}$$

for every $\sigma_{ij} = \sigma_{ji}$ and positive constant m and M, the following equivalence relation is valid:

$$m\{\int_\Omega \sigma_{ij}(\underset{\sim}{w})\sigma_{ij}(\underset{\sim}{w})d\Omega\}^{1/2} \leq [a(\underset{\sim}{w},\underset{\sim}{w})]^{1/2} \leq M\{\int_\Omega \sigma_{ij}(\underset{\sim}{w})\sigma_{ij}(\underset{\sim}{w})d\Omega\}^{1/2} \tag{19}$$

This means that the relation (15) can be obtained from the inequality (5). Therefore, it is possible to define the mean square norm of the stress tensor (14) as the error measure of finite element approximations. Stresses are often primary quantities to be obtained by finite element approximations, especially for the design of structures. Thus, the error measure (14) for the stress tensor is meaningful.

Another interesting measure of the approximation error is defined by the mean square of the displacement:

$$e_h = \{\int_\Omega(\underset{\sim}{u}-\underset{\sim}{u}_h)\cdot(\underset{\sim}{u}-\underset{\sim}{u}_h)d\Omega\}^{1/2} \tag{20}$$

In this case, for the elastic body considered according to Korn's inequality (see e.g., Nečas and Hlaveček [21]) there is a positive constant $\bar{m}$ such that

$$\bar{m}\{\int_\Omega \underset{\sim}{w}\cdot\underset{\sim}{w}\, d\Omega\}^{1/2} \leq a(\underset{\sim}{w},\underset{\sim}{w})^{1/2}, \tag{21}$$

for every $\underset{\sim}{w}$ which is not exactly a rigid body motion, that is,

$$\underset{\sim}{w}(\underset{\sim}{x}) \neq \underset{\sim}{a} + \underset{\sim}{b} \times \underset{\sim}{x} \tag{22}$$

for constant vectors $\underset{\sim}{a}$ and $\underset{\sim}{b}$. Thus if the inequality

$$(u_i-v_{hi})_{,j}(u_i-v_{hi})_{,j} \leq \hat{M}(u_i-v_{hi})(u_i-v_{hi}) \tag{23}$$

is assumed for the solution $\underset{\sim}{u}$ of (4) and its interpolation $\underset{\sim}{v}_h$, it is possible to find a positive constant $\bar{M}$ such that

$$a(\underset{\sim}{u}-\underset{\sim}{v}_h,\ \underset{\sim}{u}-\underset{\sim}{v}_h)^{1/2} \leq \bar{M}\{\int_\Omega(\underset{\sim}{u}-\underset{\sim}{v}_h)\cdot(\underset{\sim}{u}-\underset{\sim}{v}_h)d\Omega\}^{1/2} \tag{24}$$

Then the quantity $a(\underset{\sim}{w},\underset{\sim}{w})$ is equivalent to the mean square displacement, that is, (20) can be used as an error measure of finite element approximations. Although (21) can be verified by Korn's inequality, the assumption (23) does not always hold. In fact, for a one-dimensional problem, if

$$u-v_h = a \sin n\pi x$$

is assumed, the constant $\hat{M}$ goes to infinity as $n \to \infty$. This means that the existence of $\hat{M}$ cannot be assumed for an arbitrary function $\underset{\sim}{u}$. Despite this fact, let us assume that the solution $\underset{\sim}{u}$ of (4) and its interpolation $\underset{\sim}{v}_h$ satisfy inequality (23). Then it is possible to imply a positive constant C independent of discretization such that

$$C\, e_h \leq e_h^I \overset{\Delta}{=} \{\int_\Omega (\underset{\sim}{u}-\underset{\sim}{v}_h)\cdot(\underset{\sim}{u}-\underset{\sim}{v}_h)\}^{1/2} \tag{25}$$

Thus far we have introduced three different error measures e_h, that is, three different error measures E_e for the grid optimization, namely:

$$E_e = \begin{cases} \{a_e(\underset{\sim}{u}-\underset{\sim}{v}_h,\ \underset{\sim}{u}-\underset{\sim}{v}_h)\}^{1/2} \\ \{\int_{\Omega_e} \sigma_{ij}(\underset{\sim}{u}-\underset{\sim}{v}_h)\sigma_{ij}(\underset{\sim}{u}-\underset{\sim}{v}_h)d\Omega\}^{1/2} \\ \{\int_{\Omega_e} (\underset{\sim}{u}-\underset{\sim}{v}_h)\cdot(\underset{\sim}{u}-\underset{\sim}{v}_h)d\Omega\}^{1/2} \end{cases} \tag{26}$$

For these choces, the finite element grid design problem (1) means that the error in the strain energy, the stress tensor, and the displacement vector is "minimized," respectively.

We mention two other possibly useful measures of error for the grid design problem:

$$E_e = \begin{cases} \{\int_{\Omega_e} \bar{\sigma}(\underset{\sim}{u}-\underset{\sim}{v}_h)^2 d\Omega\}^{1/2} \\ \{\int_{\Omega_e} \tau_{max}(u-v_h)^2 d\Omega\}^{1/2} \end{cases} \tag{27}$$

where $\bar{\sigma}$ and τ_{max} are the equivalent (von Mises) and the maximum shear stresses, defined by

$$\bar{\sigma} = \left(\frac{3}{2}\sigma_{ij}^D\sigma_{ij}^D\right)^{1/2}$$

$$\tau_{max} = \mathrm{Max}\{|\sigma_1-\sigma_2|/2,\ \ |\sigma_2-\sigma_3|/2,\ \ |\sigma_3-\sigma_1|/2\}$$

respectively. Here $\underset{\sim}{\sigma}^D$ represents the deviatoric stress tensor, and $\{\sigma_1,\sigma_2,\sigma_3\}$ are the principal sgresses of $\underset{\sim}{\sigma}$.

Approximations of the Necessary Condition

As shown in the previous sections, error measures E_e for the grid design problem (1) are defined by the difference between the solution $\underset{\sim}{u}$ of the problem (4) and its interpolation $\underset{\sim}{v}_h$. Now we shall consider approximations of a quantity E_e using the finite element solution $\underset{\sim}{u}_h$ which can be obtained by computations. The procedure is described using the one-dimensional problem which is characterized by the bilinear form

$$a(u,v) = \int_0^L EA\, u'v'\, dx \tag{28}$$

where EA is the axial rigidity of a bar and u' is the first derivative of axial displacement u. Suppose that u and v are approximated by piecewise linear polynomials. In this case, for the error measure

$$E_e = \{a_e(u-v_h,\ u-v_h)\}^{1/2}$$

the following estimate holds:

$$E_e = \{\int_{\Omega_e} EA(u-v_h)'(u-v_h)'dx\}^{1/2} \le h_e \{\int_{\Omega_e} |EA\ u''(s)|^2 d\Omega\}^{1/2} \tag{29}$$

Here the rigidity EA is assumed to be constant in each element Ω_e. This estimate can be obtained by using Taylor's series for u' and applying the Schwarz inequality. Indeed

$$\begin{aligned}(u'-v_h')(x) &= \frac{1}{h_e}\{\int_{x_{e-1}}^{x} (s-x_{e-1})u''(s)ds - \int_{x_e}^{x} (x_e-s)u''(s)ds\} \\ &\le \int_{x_{e-1}}^{x_e} |u''(s)|ds \le \sqrt{h_e}^{\,1/2} \{\int_{x_{e-1}}^{x_e} |u''(s)|^2 ds\}^{1/2},\end{aligned} \tag{30}$$

where $h_e = x_e - x_{e-1}$. Then we have

$$\int_{\Omega_e} EA|u'-v_h|(x)^2 dx \le h_e^2 \int_{\Omega_e} |EA\ u''(s)|^2 ds.$$

Thus, the error measure E_e for the design of the finite element grid is bounded by

$$E_e \le \hat{E}_e \triangleq h_e \{\int_{\Omega_e} EA|u''(s)|^2 ds\}^{1/2}. \tag{31}$$

It is noted that the above upper-bound estimate, called an estimate of interpolation error, has been studied thoroughly for functions in Sobolev spaces $H^m(\Omega)$ defined on multi-dimensional domain Ω (see, e.g., Ciarlet [22]).

Now the only problem remaining is to approximate the "second" derivative of the solution u which is unknown *a priori*. Our task is to construct a function $\hat{u}$ from the known quantity u_h of the finite element solution. Although there are infinitely many ways to construct a function $\hat{u}$, we follow the idea suggested by Nakazawa and Zienkiewicz [23]. To explain the idea, let us suppose that Ω_e is a line element consisting of two nodes which yields a piecewise linear approximation of u in Ω. Then in each element Ω_e a constant first derivative of the finite element solution u_h can be computed. Let us identify this as the value of the first derivative of u_h at the centroid of Ω_e, and let us denote this by τ_e. Then the first derivative of u_h is expressed by $u_h'(x) = \sum_{e=1}^{E} \tau_e H_e(x)$ in Ω, where $H_e(x) = 1$ if $x \in \Omega_e$ and $H_e(x)=0$ if $x \notin \Omega_e$. Now we shall construct a continuous function $\hat{u}'$ using the same interpolation functions for the displacement u_h, that is, $\hat{u}'(x) = \sum_{\alpha=1}^{2} u_\alpha' N_\alpha(x)$ in Ω_e, where $\{N_\alpha(x)\}$ are the shape functions. The coefficients $\hat{u}_\alpha'$ are computed by the method of least squares:

$$\underset{\hat{u}_\alpha'}{\text{Min}}\ (1/2 \int_\Omega (\hat{u}'(x) - u_h'(x))^2 d\Omega). \tag{32}$$

This yields the necessary condition

$$\sum_{e=1}^{E} (\int_{\Omega_e} N_\alpha N_\beta d\Omega)\hat{u}_\beta' = \sum_{e=1}^{E} (\int_{\Omega_e} N_\alpha d\Omega)\tau_e \tag{33}$$

Solving this system of linear equations, $\hat{u}'$ can be obtained. Since $\hat{u}'$ is now a piecewise linear continuous function, it possesses the first derivative of $\hat{u}'$, i.e., the second derivative of $\hat{u}$. Using this $\hat{u}$, we approximate the upper-bound of the error measure E_e by

$$\hat{E}_e \doteq E_e^h \triangleq h_e\{\int_{\Omega_e} EA|\hat{u}''(s)|^2 ds\}^{1/2}. \tag{34}$$

In the above, we have introduced two approximations of the error measure E_e for the finite element grid design, that is,

$$E_e \leq \hat{E}_e \doteq E_e^h. \tag{35}$$

Using these we shall approximate the necessary condition (3) to the grid optimization problem (1) in the form

$$E_e^h = \text{Constant}, \qquad e=1,\ldots,E. \tag{36}$$

The quantity E_e^h is called *an error indicator* in an element Ω_e, and is applied for constructing numerical algorithms to predict the optimal grid.

The exact same procedure is applicable to multi-dimensional problems. Indeed, if plane linearly elastic structures are considered, the bilinear form $a(\underset{\sim}{u},\underset{\sim}{v})$ is defined by

$$a(\underset{\sim}{u},\underset{\sim}{v}) = \int_\Omega E_{ijk\ell}u_{k,\ell}v_{i,j}\, d\Omega \tag{37}$$

where $E_{ijk\ell}$ is the inverse of the compliance matrix $C_{ijk\ell}$ and satisfies the symmetry condition

$$E_{ijk\ell} = E_{ij\ell k} = E_{jik\ell} = E_{k\ell ij}. \tag{38}$$

In this case, one of the key ideas is to use isoparametric transformations to evaluate $a_e(\underset{\sim}{u},\underset{\sim}{v})$. Suppose that an element Ω_e in the (x_1,x_2) coordinate system is constructed by the square master element in the normalized coordinate system (ξ_1,ξ_2). Then the isoparametric transformation is assumed as

$$x_i = x_{i\alpha}N_\alpha(\underset{\sim}{\xi})$$

where $N_\alpha(\underset{\sim}{\xi})$, $\alpha=1,\ldots,N_e$, are the shape functions in the master element, and N_e is the number of nodes in Ω_e. Then defining the Jacobian matrix $\underset{\sim}{J}$ by

$$J_{ij} = \frac{\partial x_j}{\partial \xi_i} = x_{j\alpha}\frac{\partial N_\alpha}{\partial \xi_i},$$

the first derivative in x_i is expressed by

$$\partial\phi/\partial x_i = J_{ij}^{-1}\, \partial\phi/\partial\xi_j$$

where $\underset{\sim}{J}^{-1}$ is the inverse of $\underset{\sim}{J}$. Using these results we have

$$\begin{aligned} a_e(\underset{\sim}{u},\underset{\sim}{v}) &= \int_{\Omega_e} E_{ijk\ell}(\partial u_k/\partial x_\ell)(\partial v_i/\partial x_j)d\Omega \\ &= \int_{-1}^{1}\int_{-1}^{1} E_{ijk\ell}J_{\ell q}^{-1}J_{jp}^{-1}(\partial u_k/\partial\xi_q)(\partial v_i/\partial\xi_p)J d\xi_1 d\xi_2. \end{aligned} \tag{39}$$

where J is the determinant of $\underset{\sim}{J}$. For $N_e = 4$, applying the inequality similar to

(30), we have

$$\frac{\partial}{\partial \xi_q}(u_k - v_{hk})(\xi) \leq \int_{-1}^{1} \left|\frac{\partial^2}{\partial \xi_q \partial \xi_q} u_k\right| d\xi_q \leq \{\int_{-1}^{1}\int_{-1}^{1} \left|\frac{\partial^2}{\partial \xi_q \partial \xi_q} u_k\right|^2 d\xi_1 d\xi_2\}^{1/2}$$

on $\xi_p = \pm 1$, $p \neq q$, where no summation is taken in q. Thus, we have

$$a_e(\underset{\sim}{u}-\underset{\sim}{v}_h, \underset{\sim}{u}-\underset{\sim}{v}_h) \leq \{\int_{-1}^{1}\int_{-1}^{1} |E_{ijk\ell} J^{-1}_{\ell q} J^{-1}_{jp}| \, J d\xi_1 d\xi_2\}$$
$$\cdot \{\int_{-1}^{1}\int_{-1}^{1} \left|\frac{\partial}{\partial \xi_q}\left(\sum_{s=1}^{2} \frac{\partial u_k}{\partial \xi_s}\right)\right|^2 d\xi_1 d\xi_2\}^{1/2} \tag{40}$$
$$\cdot \{\int_{-1}^{1}\int_{-1}^{1} \left|\frac{\partial}{\partial \xi_p}\left(\sum_{r=1}^{2} \frac{\partial u_i}{\partial \xi_r}\right)\right|^2 d\xi_1 d\xi_2\}^{1/2}$$

This yields the upper-bound of the interpolation error E_e such that

$$a_e(\underset{\sim}{u}-\underset{\sim}{v}_h, \underset{\sim}{u}-\underset{\sim}{v}_h) \leq (\hat{E}_e)^2 \triangleq \{\int_{-1}^{1}\int_{-1}^{1} |E_{ijk\ell} J^{-1}_{\ell q} J^{-1}_{jp}| \, J d\xi_1 d\xi_2\}$$
$$\cdot \{\int_{-1}^{1}\int_{-1}^{1} \left|\frac{\partial}{\partial \xi_q}\left(\sum_{s=1}^{2} \frac{\partial u_k}{\partial \xi_s}\right)\right|^2 d\xi_1 d\xi_2\}^{1/2} \tag{41}$$
$$\cdot \{\int_{-1}^{1}\int_{-1}^{1} \left|\frac{\partial}{\partial \xi_p}\left(\sum_{r=1}^{2} \frac{\partial u_i}{\partial \xi_r}\right)\right|^2 d\xi_1 d\xi_2\}^{1/2}$$

The construction of $\hat{\underset{\sim}{u}}$ from the finite element solution $\underset{\sim}{u}_h$ is exactly the same as was done for the one-dimensional problem using the least squares method (32). That is, a continuous first derivative $\partial \hat{u}_i / \partial x_j$ is obtained by (32) using the finite element solution $\partial u_{hi} / \partial x_j$. This should be very straightforward. Thus an approximated necessary condition (3) for the grid optimization problem (1) is given by

$$E_e^h = \text{Constant}, \qquad e=1,\ldots,E, \tag{42}$$

where

$$a_e(\hat{\underset{\sim}{u}}-\underset{\sim}{v}_h, \hat{\underset{\sim}{u}}-\underset{\sim}{v}_h) \leq (E_e^h)^2 \triangleq \{\int_{-1}^{1}\int_{-1}^{1} |E_{ijk\ell} J^{-1}_{\ell q} J^{-1}_{jp}| \, J d\xi_1 d\xi_2\}$$
$$\cdot \{\int_{-1}^{1}\int_{-1}^{1} \left|\frac{\partial}{\partial \xi_q}\left(\sum_{s=1}^{2} \frac{\partial u_k}{\partial \xi_s}\right)\right|^2 d\xi_1 d\xi_2\}^{1/2}$$
$$\cdot \{\int_{-1}^{1}\int_{-1}^{1} \frac{\partial}{\partial \xi_p}\left(\sum_{r=1}^{2} \frac{\partial u_i}{\partial \xi_r}\right)\Big|^2 d\xi_1 d\xi_2\}^{1/2}$$

In the above discussion, the interpolation error $a_e(\underset{\sim}{u}-\underset{\sim}{v}_h, \underset{\sim}{u}-\underset{\sim}{v}_h)$ is bounded by some upper bound using the second derivatives of the solution $\underset{\sim}{u}$. It is noted, however, that the upper bound given in (40) may not be the optimum. In order to avoid the upper bound estimation of the interpolation error $a_e(\underset{\sim}{u}-\underset{\sim}{v}_h, \underset{\sim}{u}-\underset{\sim}{v}_h)$, the following identity satisfied in a 4-node bilinear isoparametric element may be applied:

$$\left(\frac{\partial u_k}{\partial \xi_q} - \frac{\partial v_{hk}}{\partial \xi_q}\right)(\xi_1,\xi_2) = \frac{1}{4}(1+\xi_p)\int_{\xi_p}^{1}\left\{\frac{\partial u_k}{\partial \xi_p}\bigg|_{\xi_q=-1} - \frac{\partial u_k}{\partial \xi_p}\bigg|_{\xi_q=1}\right\}d\xi_p$$

$$+\frac{1}{4}(1-\xi_p)\int_{\xi_p}^{-1}\left\{\frac{\partial u_k}{\partial \xi_p}\bigg|_{\xi_q=-1} - \frac{\partial u_k}{\partial \xi_p}\bigg|_{\xi_q=1}\right\}d\xi_p \tag{44}$$

$$+\frac{1}{2}\left\{\int_{\xi_q}^{-1}\int_{\xi_q}^{-1}\frac{\partial^2 u_k}{\partial \xi_q \partial \xi_q}d\xi_q d\xi_q - \int_{\xi_q}^{1}\int_{\xi_q}^{1}\frac{\partial^2 u_k}{\partial \xi_q \partial \xi_q}d\xi_q d\xi_q\right\}$$

Using this, the error measure E_e in each finite element Ω_e can be defined by

$$E_e = \sqrt{a_e(u-v_h,\ u-v_h)}$$

$$\triangleq \left\{\int_{-1}^{1}\int_{-1}^{1} E_{ijk\ell}\sum_{p,q} J_{\ell q}^{-1}J_{jp}^{-1}\tilde{u}_{k,qq}\tilde{u}_{i,pp}\,J d\xi_1 d\xi_2\right\}^{1/2} \tag{45}$$

where

$$\tilde{u}_{k,qq} = \text{right hand side of (44).}$$

It is clear that the definition (45) should be sharper than (41), although the evaluation of identity (44) is somewhat complicated.

If the error measure defined by the stress tensor

$$E_e = \left\{\int_{\Omega_e}\sigma_{ij}(\underset{\sim}{u} - \underset{\sim}{v}_h)\sigma_{ij}(\underset{\sim}{u} - \underset{\sim}{v}_h)d\Omega\right\}^{1/2} \tag{46}$$

is taken as the basis for grid optimization, a derivation similar to the one given above leads to the error indicator

$$E_e^h \triangleq \left\{\int_{-1}^{1}\int_{-1}^{1}\left|\frac{\partial\hat{\sigma}_{ij}}{\partial \xi_p}\frac{\partial\hat{\sigma}_{ij}}{\partial \xi_p}\right| J d\xi_1 d\xi_2\right\}^{1/2}$$

where $\hat{\underset{\sim}{\sigma}}$ is the continuous stress tensor obtained by a discontinuous stress tensor $\underset{\sim}{\sigma} = \underset{\sim}{\sigma}(\underset{\sim}{u}_h)$ using the least square method (32). Similarly, if

$$E_e = \left\{\int_{\Omega_e}(\underset{\sim}{u} - \underset{\sim}{v}_h)\cdot(\underset{\sim}{u} - \underset{\sim}{v}_h)d_\Omega\right\}^{1/2} \tag{47}$$

is taken as the error measure for the grid design problem, the corresponding error indicator may be defined by

$$E_e^h \triangleq \left\{\int_{-1}^{1}\int_{-1}^{1}\frac{\partial u_{hi}}{\partial \xi_p}\frac{\partial u_{hi}}{\partial \xi_p} J d\xi_1 d\xi_2\right\}^{1/2} \tag{48}$$

where $\underset{\sim}{u}_h$ is the finite element solution. It is clear that the error indicator (48) is similar to the strain energy in an element Ω_e.

Quality Control of Finite Element Approximations

The error indicators E_e^h, e=1,...,E, defined for finite elements are computed by finite element solution $\underset{\sim}{u}_h$ as shown above, and they are approximations of the error measures E_e, e=1,...,E. Thus, the quantities E_e^h, e=1,...,E, indicate the quality of the finite element approximation for the given grid. If E_e^h, e=1,...,E, are almost constant, then the finite element grid is almost optimal. On the other hand, if E_e^h, e=1,...,E, are far from having constant value the grid is "poor." To describe the degree of optimality of grids, let us define an indicator of quality (I.Q.) by

$$I.Q. = \max_{e=1,\dots,E} E_e^h/E_{av}^h \tag{49}$$

where E_{av}^h is the average of E_e^h, e=1,...,E;

$$E_{av}^h = \sum_{e=1}^{E} E_e^h/E. \tag{50}$$

If the finite element grid is optimal, I.Q. has value unity.

In finite element analyses, if I.Q. and E_e^h, e=1,...,E, are produced as output, then the quality of approximation can be evaluated. In other words, it is possible to determine an improved finite element model using the number I.Q. for models consisting of a like number of elements. Although the absolute magnitude of the total error is unknown, the quality of finite element approximation can still be maintained by enforcing the value of I.Q. to be close to unity.

On Computational Solution

Applying the so-called optimality criterion method, we shall construct an algorithm to obtain the optimal finite element grid system. That is, using the approximated necessary condition of the grid design problem (1), namely

$$E_e^h = \text{Constant}, \qquad e=1,\dots,E \tag{51}$$

as a basis, we shall describe a method to predict the relocation of nodes in a finite element model Ω_h. In order to explain the algorithm, plane problems are considered which are discretized by four-node "isoparametric" elements. For simplicity, suppose that the n-th node is shared by four finite elements Ω_{ni}, i=1,...,4, as shown in Fig. 1, and that the error indicators are computed as E_{ni}^h, i=1,...,4. The algorithm is designed to determine an appropriate relocation of node n such that the optimality condition (51) is met, i.e., to achieve $E_{n1}^h = E_{n2}^h = E_{n3}^h = E_{n4}^h$. Noting that the area of the element in which the error indicator E_e^h is the largest should be reduced, the following scheme for the relocation of nodes is introduced:

$$\underset{\sim}{x}_n = \sum_{i=1}^{4} \underset{\sim}{x}_{ni} \left(\frac{E_{ni}^h}{A_{ni}}\right) / \sum_{i=1}^{4} \left(\frac{E_{ni}^h}{A_{ni}}\right). \tag{52}$$

where A_{ni} and x_{ni} are the area and the coordinates of the centroids of the elements

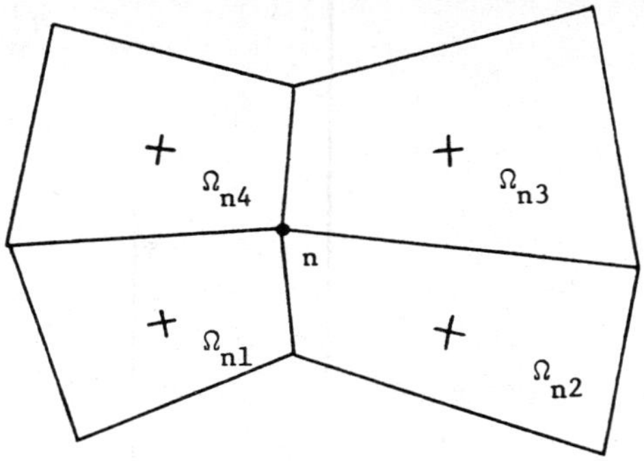

Figure 1. Typical node n and adjacent elements.

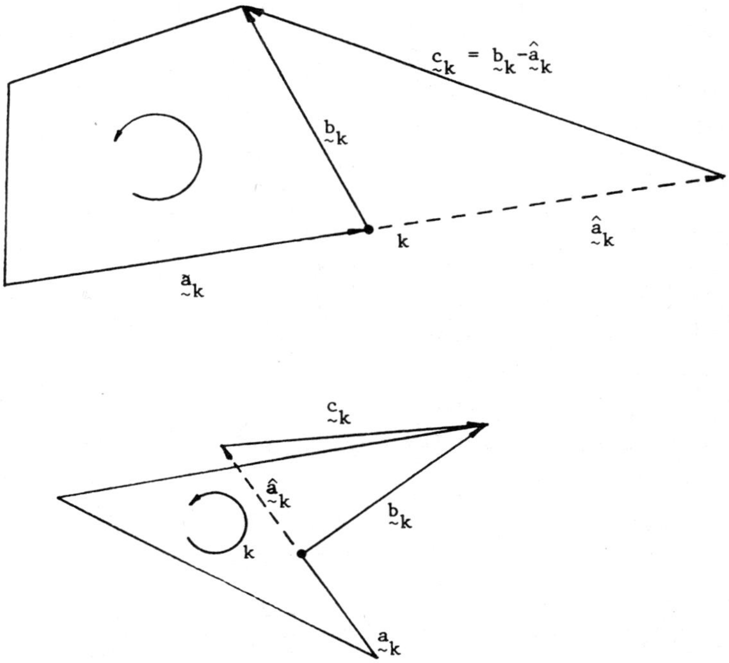

Figure 2. Check for convexity.

Ω_{ni}, i=1,...,4, respectively. This scheme implies that the new location of the n-th node becomes closer to the centroid of the element which has the largest error indicator among them. Furthermore, it is noted that if (51) is met for the rectangular elements Ω_{ni}, i=1,...,4, then the location of the n-th node is unchanged by the relocation scheme (52). Unless the weight $1/A_{ni}$ is applied to the error indicator E^h_{ni}, we cannot maintain its original location for the optimal situation. Another interesting remark on the relocation scheme (49) is that the new position of the n-th node by (49) is, in general, the domain occupied by four elements sharing it, although some exceptions may exist. In other words, while the crashing of elements is possible with the node relocation scheme (52), the occurrence of crashing is unlikely.

Where four-node elements are used there is a possibility that they become nonconvex, i.e., at certain integration points the determinant to compute element stiffness matrices using the idea of isoparametric transformations may become nonpositive. It is necessary, therefore, to have means to check for convexity of the elements at each step of node relocation.

The method applied in this study makes use of the sign of the determinant defined by two vectors $\hat{\underset{\sim}{a}}_k$ and $\underset{\sim}{c}_k$, k=1,...,4, shown in Fig. 2, where k denotes an arbitrary node number in an element. For the node k, we first define two vectors $\underset{\sim}{a}_k$ and $\underset{\sim}{b}_k$ counter-clockwise along the edges of a finite element, and then shift the vector $\underset{\sim}{a}_k$ to the node k in order to define the new vector $\hat{\underset{\sim}{a}}_k$. The last vector $\underset{\sim}{c}_k$ is defined by the difference of $\underset{\sim}{b}_k$ and $\hat{\underset{\sim}{a}}_k$. If the determinant d_k at the node k is given by

$$d_k = \begin{vmatrix} \hat{a}_{kx} & \hat{a}_{ky} \\ c_{kx} & c_{ky} \end{vmatrix} \tag{53}$$

where $\hat{\underset{\sim}{a}}_k = \hat{a}_{kx} \underset{\sim}{i} + \hat{a}_{ky} \underset{\sim}{j}$ and $\underset{\sim}{c}_k = c_{kx} \underset{\sim}{i} + c_{ky} \underset{\sim}{j}$, convexity of the element can be determined by its sign. Indeed, if $d_k < 0$, then the element is nonconvex at the node k. If $d_k > 0$, the element maintains convexity at the node k. If $d_k \leq 0$ is computed for the new location of node, we have to "freeze" the node to avoid nonconvexity of elements. In other words, we skip relocation of this particular node.

As a practical point, we note that relocation of nodes can be performed several times using a single set of values for error indicators E^h_e computed for a certain finite element grid system. This obviously accelerates the iterative scheme using (52) to achieve the optimality (51).

It is noted that the relocation scheme (52) is a particular choice applied in the present study. There are infinitely many ways to relocate nodes.

A typical flow chart for grid optimization is as follows:

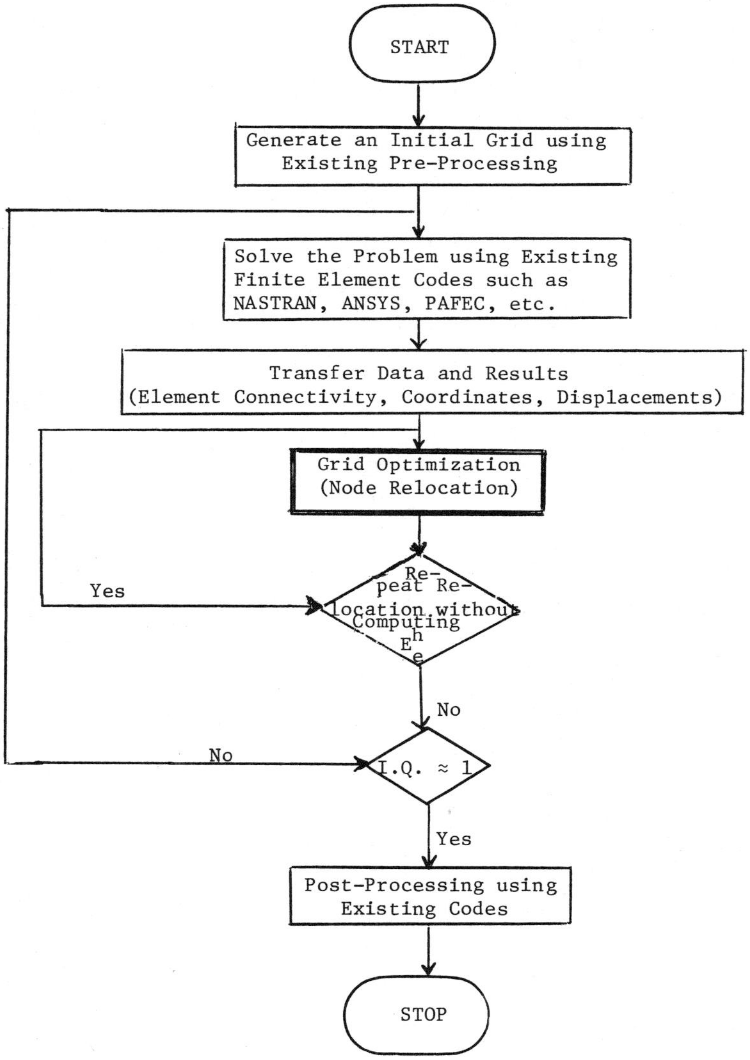

Grid optimization may be performed as a separate step, independent of the means for finite element analysis. In other words, the grid optimization can be incorporated into a systems that makes use of existing FEM codes.

Numerical Examples

Three examples are described here. The first two are plane elasticity problems. The third is obtained from flow analysis governed by Navier-Stokes' equations.

The first example represents stress analysis of a thin plate with notches subjected to inplane tension field as shown in Fig. 3. Starting from the uniform grid given in Fig. 4, an improved grid is obtained as shown in Fig. 5 using the error indicator E_e^h defined by the bilinear form $a(\cdot,\cdot)$. We assume isotropy and homogeneity of the material, and values for the material constants as follows:

Young's Modulus: $E = 1.02 \times 10^5$ MPa

Poisson's Ratio: $\nu = 0.3$.

The thickness of the plate is $t = 0.024$m, and the applied tension is $\sigma_o = 600$ MPa along the top surface. Using symmetry, only a quarter part of the notched plate is discretized as shown in Fig. 4. At the fifth iteration, the ratio of the arithmetic and geometric means of the deviation of the error indicators E_e^h, $e=1,\ldots,E$, becomes less than two. The distribution of the stress component ${}_{yy}$ along the x axis is given in Fig. 6. It is clear that the singularity of the stress near to the tip of the notch is simulated very well by the improved finite element grid, although convergence of the numerical algorithm to satisfy the necessary condition in the exact sense is not obtained yet. An interesting remark is that the total strain energy of the thin plate is almost unchanged during the finite element grid modification. Indeed, it is 0.2608×10^{-2} for the initial uniform grid, and is 0.2636×10^{-2} for the final form. Only 1% difference can be observed in the whole process, while a significant change is obtained for the distribution of the stresses near to the top of the notch.

The second example is again for stress analysis, but for a structure which is nonhomogeneous. There is a hole inside of a foundation, packed with a soft material, as shown in Fig. 7. Material constants are assumed as

(Foundation)

Young's Modulus $E = 6. \times 10^{4}$ MPa

Poisson's Ratio $\nu = 0.23$

(Hole)

Young's Modulus $E = 1.45 \times 10^{3}$ MPa

Poisson's Ratio $\nu = 0.16$

Assuming plane strain, the thickness is unity. The constant body force $f_y = -0.00258$ MN/m^3 is assumed in the downward vertical directions, and the uniform traction $t_y = -0.0262$ MPa on the top surface is applied in the downward vertical direction. The relocation process of nodes is applied four times from the uniform grid given in Fig. 8, and provides an improved finite element grid shown in Fig. 9. The ratio of the arithmetic and geometric means of the deviation of the error indicators is 1.83510 at the fourth iteration while it was 3.0209 at the initial. During the iteration process, the total strain energy increased only 0.4%. Figure 10 shows the distribution of the maximum shear stress τ_{max} along the wall of the hole. It is clear that "singular" behavior of the stress at corner points, is again controlled by the improved grid obtained via the node relation algorithm.

The last example is related to flow analysis governed by the Navier-Stokes equation. Since this is a nonlinear problem, application of the node relocation method is not the same as for linear problems. In the present example, the following strategy is taken. Applying the Newton-Raphson method, the system of nonlinear equations is solved. In this case, up to a certain number of iterations we have applied the relocation algorithm after each solution by the Newton-Raphson method. When the relative error of the Newton-Raphson method becomes small enough, the node relocation algorithm is abandoned in order to obtain the convergent result. For the flow problem described in Fig. 11, the relocation algorithm is applied for the first four iterations, starting from the initial grid given in Fig. 12. The velocity itself is used as the error-measure for this problem. The improved grid and the velocity field computed using this grid are shown in Fig. 13.

Acknowledgement

The work reported in this paper received partial support under the National Science Foundation grant CEE 8118158 and also partial support from the National Aeronautics and Space Administration through the grant NAG3-388.

References

[1] Babuška, I., J. Chandra and J. E. Flaherty (eds), Proc. Symp. on Adaptive Comp. Methods for Partial Differential Equations, SIAM, Philadelphia, 1983.

[2] Ghia, K. and Ghia, U., Advances in Grid Generation, FED - Vol. 5, American Society of Mechanical Engineers, New York, 1983.

[3] Oliveira, E. R. de A., "Optimization of finite element solution," Proc. 3rd Conf. Matrix Methods in Structural Mechanics, Wright-Patterson Air Force Base, Ohio, 1971.

[4] Prager, William, "A note on the optimal choice of finite element grids," Comp. Meths. Appld. Mech. & Engrg., 6, 363-366, 1975.

[5] Tang, J. W. and Turcke, "Characteristics of optimal grids," Compt. Method. Appl. Mech. Eng., 11, 31-37, 1977.

[6] Masur, E. F., "Some remarks on the optimal choice of finite element grids," Comp. Meths. Appl. Mechs. & Engrg., 14, 237-248, 1978.

[7] Melosh, R. J., and Marcal, P. V., "An energy basis for mesh refinement of structural continua," Int. J. Num. Meth. Engrg., 11, 1083-1091, 1977.

[8] Turcke, D. J. and McNeice, G. M., "Guidelines for selecting finite element grids based on an optimization study," Computers & Structures, 4, 499-519, 1974.

[9] Shephard, M. S., Gallagher, R. H. and Abel, J. F., "The synthesis of near-optimum finite element meshes with iterative computer graphics," Int. J. Num. Meth. Engrg., 15, 1021-1039, 1980.

[10] Babuska, I. and Rheinboldt, W. C., "Reliable error estimation and mesh adaptation for the finite element method," in J. T. Oden, Ed., Computational Methods in Nonlinear Mechanics, North-Holland, Amsterdam, 67-108, 1980.

[11] Babuska, I. and Rheinboldt, W. C., "Error estimates for adaptive finite element computations," SIAM J. Num. Anal., 15, 736-754, 1978.

[12] Babuska, I. and Rheinboldt, W. C., "Computer error estimates and adaptive processes for some nonlinear structural problems," Comp. Meth. Applied Mech. Engrg., 34, 895-937, 1982.

[13] Babuska, I. and Szabo, B., "On the rates of convergence of the finite element methods," Int. J. Num. Meth. Engrg., 18, 323-341, 1982.

[14] Sewell, G., "An adaptive computer program for the solution of DIV(P(x,y)GRAD U) = F(x,y,U) on a polygonal region," in J. R. Whiteman, Ed., The Mathematics of Finite Elements and Applications, II, MAFGLAP, 1975; Academic Press, London, 1976.

[15] Zienkiewicz, O. C., S. R. Gago, J. P. de, and Kelly, D. W., "The hierarchical concept in finite element analysis," Computers & Structures, 16, 53-65, 1983.

[16] Felippa, C. A., "Optimization of finite element grids by direct energy search," Appl. Math. Modelling, 1, 93-96, 1976.

[17] Felippa, C. A., "Numerical experiments in finite element grid optimization by direct energy search," Appl. Math. Modelling, 1, 239-244, 1977.

[18] Shephard, M. S., "Finite element grid optimization," ASME Special Publication PVP-38, AISME, New York, 1979.

[19] Rossow, M. P. and Katz, I. N., "Hierarchical finite elements and precomputed arrays," Int. J. Num. Meth. Engrg., 12, 977-999, 1978.

[20] Díaz Bermúdez, Alejandro R., "Optimization of finite element grids using interpolation error," Ph.D. Thesis in Aerospace Engineering, The University of Michigan, Ann Arbor, 1982.

[21] Nečas, J. and Hlaváček, L., Mathematical Theory of Elastic and Elastic-Plastic Bodies, Elsevier, Amsterdam, 1981.

[22] Ciarlet, P. G., Numerical Analysis of the Finite Element Methods, Les Presses de L'Université de Montréal, 1976.

[23] Nakazawa, S., and O. Zienkiewicz, (private communication).

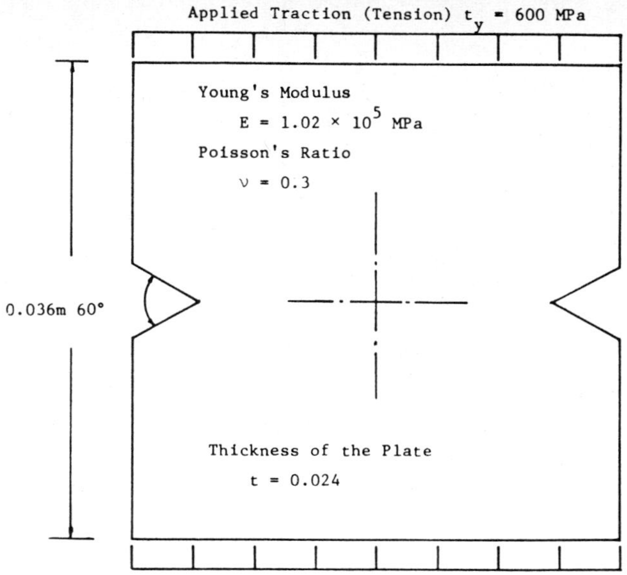

Figure 3. Notched thin plate.

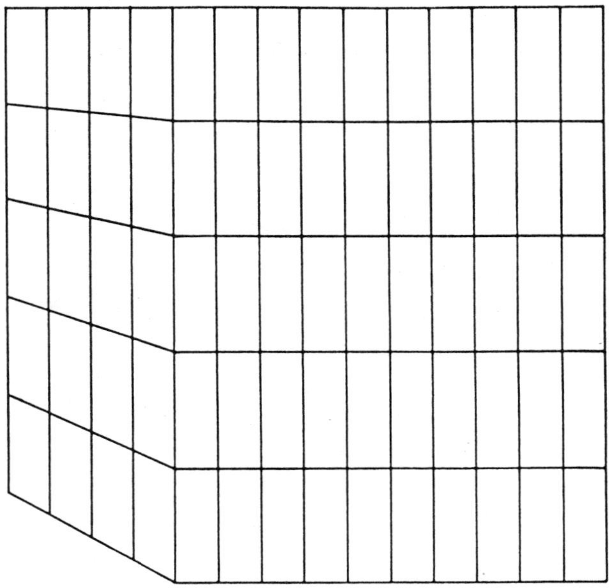

Figure 4. Initial uniform finite element grid.

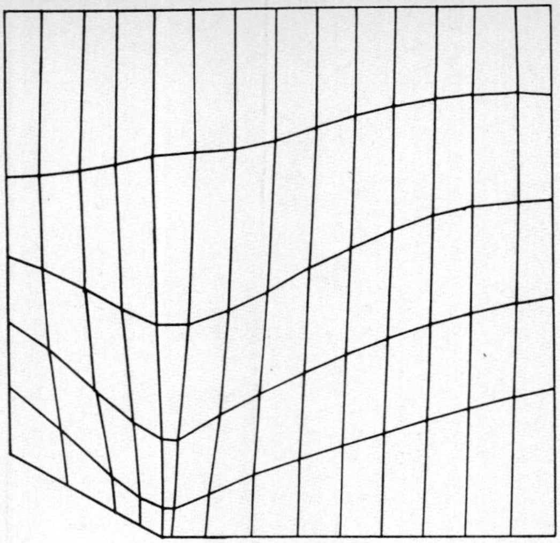

Figure 5. Improved finite element grid.

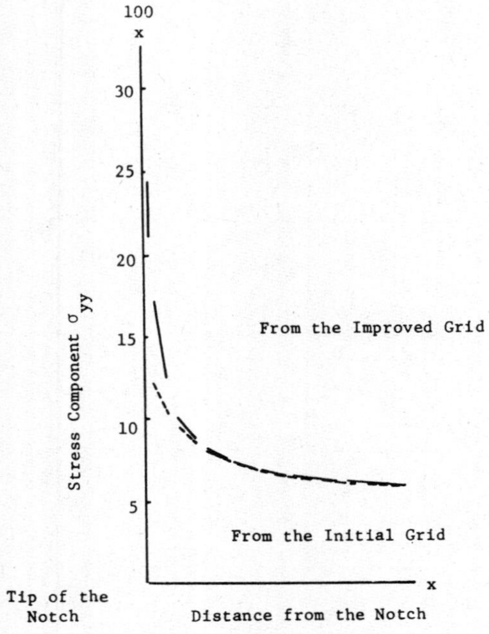

Figure 6. Stress distributions along the x-axis.

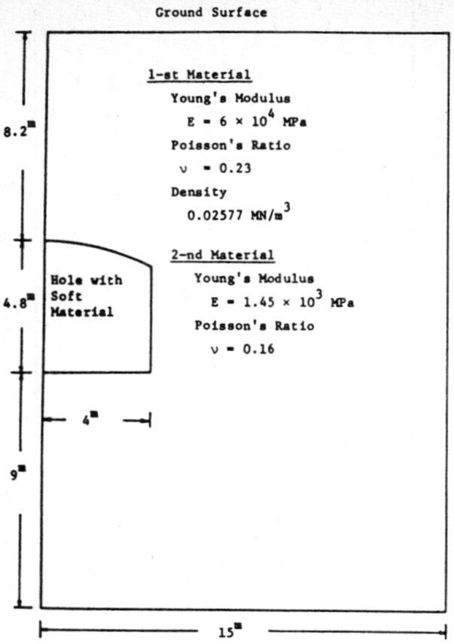

Figure 7. Stress analysis of foundation.

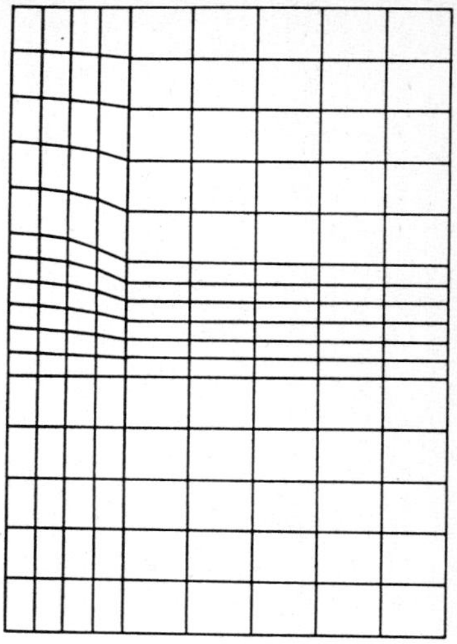

Figure 8. Initial finite element grid for foundation example problem.

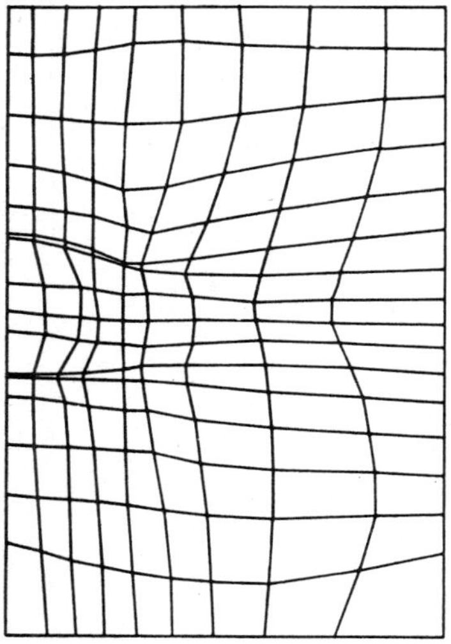

Figure 9. Improved finite element grid for foundation example problem.

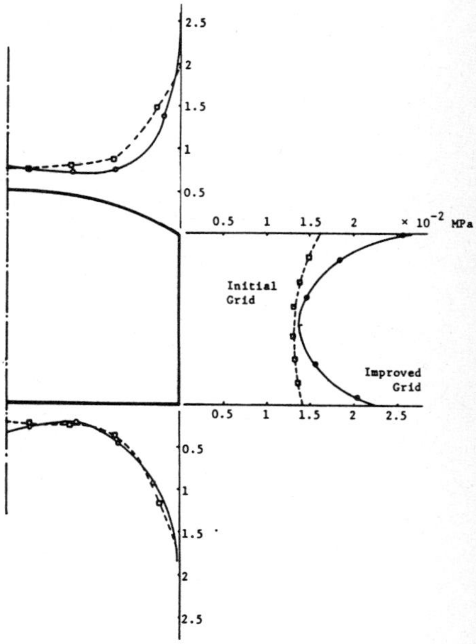

Figure 10. Distribution of the maximum shear stress τ_{max} around the hole.

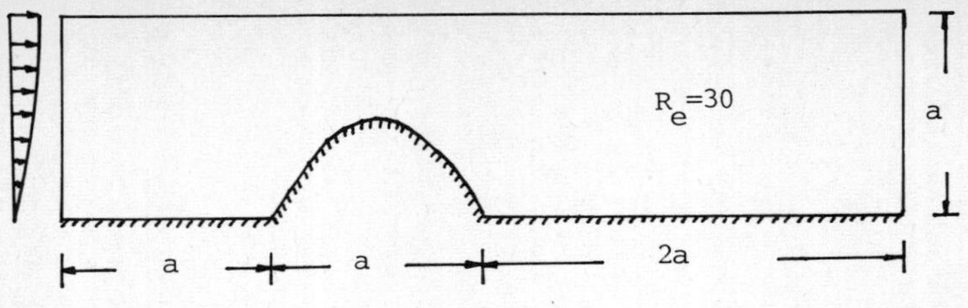

Figure 11. Navier-Stokes flow problem.

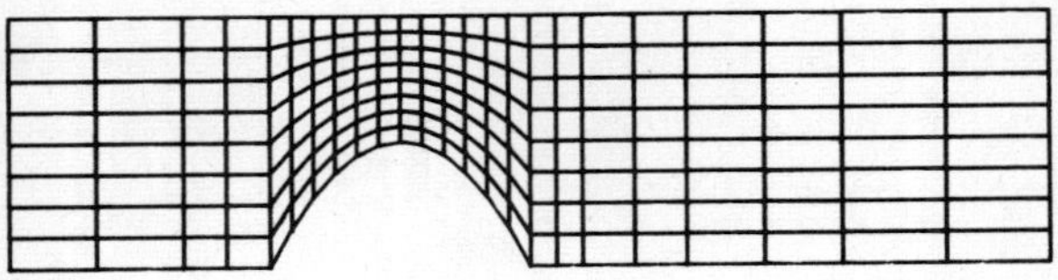

Figure 12. Initial finite element grid for Navier-Stokes flow problem.

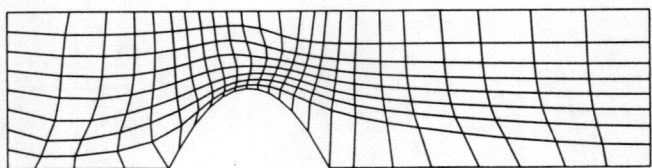

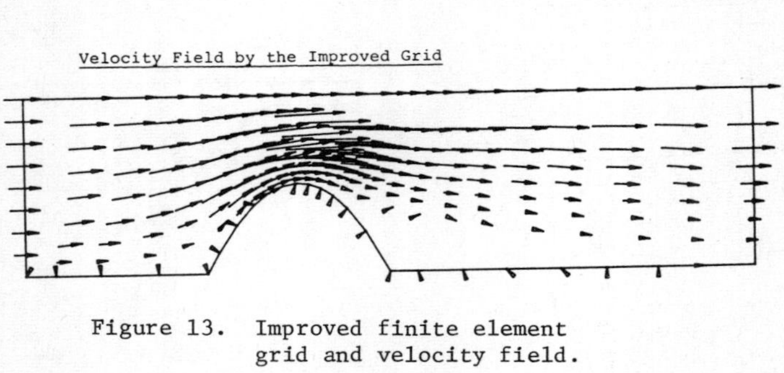

Figure 13. Improved finite element grid and velocity field.

SENSITIVITY METHODS FOR MATHEMATICAL MODELLING

Herschel Rabitz
Princeton University
Department of Chemistry
Princeton, New Jersey 08544

This paper discusses the role of sensitivity analysis in mathematical modelling with particular emphasis on differential equation models. The sensitivity analysis will be approached primarily from a functional derivative point of view for both initial and boundary value type problems. Specific illustrations will be taken from chemical kinetics involving non-linear differential equations and quantum mechanical scattering theory. Manipulation of the functional derivatives by generalized Legendre type transformations will be discussed as a means of interchanging the normal role of the systems dependent and independent variables.

I. Introduction

Mathematical modelling is becoming an increasingly pervasive endeavor in numerous areas of science and engineering. This approach to the understanding and design of physical processes has largely been stimulated in recent years by the availability of high-speed digital computers for numerically implementing the mathematical models. A serious question in a study such as this concerns the relationship between the output "observables" and the underlying input information. Quite often the mathematical models are expressed as differential equations and the input information resides as coefficients or functions in the body of the equation or in the external initial or boundary conditions. Throughout this paper all such input information will be referred to as parameters regardless of whether they are distributed in space and/or time. Some of these parameters may be controlled in the "laboratory" while others may be of a more inherent nature; in the latter case the frequent problem is lack of detailed knowledge about the precise value of the parameters. Regardless of the problem an important question concerns the relationship of the solution of the mathematical model to the underlying parameters. This question falls into the domain of sensitivity analysis and can be addressed by several methods[1]. As a natural art of "tuning up" the model many calculations may actually be performed with different parameters values. This can provide useful information related to sensitivity questions, but this brute force approach does not substitute for a more systematic and quantitative analysis based on gradient techniques. This latter approach is the one discussed in this paper. Considering $u_i(x,t)$ to be the i-th component of the solution vector of the mathematical model and α_j as the j-th constant system parameter we can define the systematic sensitivity gradients as $\frac{\partial u_i}{\partial \alpha_j}$, $\frac{\partial^2 u_i}{\partial \alpha_j \partial \alpha_k}$, etc. These partial derivatives are only applicable in circumstances where the parameters are truly constant and a more general approach allowing for spatially and temporally distributed parameters is based on functional derivatives[2]. Considering $F[g]$ to be a functional of $g(x,t)$, the first order functional derivative $\delta F/\delta g$ is defined by

$$F[g+\delta g] - F[g] = \iint \left(\frac{\delta F}{\delta g(x,t)} \right) \delta g(x,t)\,dx\,dt + \cdots \tag{1}$$

We assume that the functional derivative defined this way exists for all the physical problems of concern at least in this paper. A somewhat different but parallel development can also be based on Gâteux differentials[1,3]. In the simple case in which the parameters of the system are constant it may be shown[4] that the functional derivatives reduce to the partial derivatives under the following operation

$$\frac{\partial u_i(x,t)}{\partial \alpha_j} = \iint \frac{\delta u_i(x,t)}{\delta \alpha_j(x',t')}\,dx'dt' \tag{2}$$

where the integral is over the relevant space domain with the time restriction $0 \leq t' \leq t$. The relation in Eq. (2) clearly shows the more general nature of the functional derivatives even in the case when the nominal value of the system parameters are constant, and for this reason we shall exclusively focus on functional rather than partial derivatives. In general, one may interpret the functional derivative $\frac{\delta r(x,t)}{\delta v(x',t')}$ as the response of a function $r(x,t)$ are point x and time t with respect to a variation of $v(x',t')$ at point x' and prior time t'. The functional derivatives may be thought of as densities in space and time due to their role in the integral relations of Eqs. (1) and (2).

Although mathematical models can come in a variety of forms we shall confine ourselves here to the consideration of local differential equations

$$\underline{L}(\underline{u}(x,t),\underline{\alpha}(x,t)) = 0 \tag{3}$$

where L may be a non-linear operator with respect to the dependent solution vector $\underline{u}$ and the vector $\underline{\alpha}$ denotes the system variables residing in the differential equation. The model will be uniquely specified by the presentation of appropriate initial conditions and boundary conditions. The first order sensitivity equations are obtained by taking a variation of Eq. (3) to produce

$$\sum_j \left(\frac{\partial L_i}{\partial u_j}\right) \frac{\delta u_j(x,t)}{\delta \alpha_\ell(x',t')} + \frac{\delta L_i(x,t')}{\delta \alpha_\ell(x',t')} = 0 \qquad (4)$$

The appropriate boundary conditions and initial conditions for this equation are obtained by taking a similar variation of those conditions for Eq. (3). In this regard one can also allow for variations with respect to the parameters residing in the boundary condition of Eq. (3). Equation (4) represents a set of linear inhomogeneous differential equations for the desired sensitivity density. The differential operator $\frac{\partial L_i}{\partial u_j}$ may explicitly contain the solution to Eq. (3) which is assumed to have been solved by some suitable procedure. Since Eq. (4) is linear, it is immediately amenable to a Green's function type of analysis entailing the generation of the inverse of the differential operator involved. For a purely temporal system[5], this approach provides a practical means of solving for the system sensitivities while in boundary value problems this may not be the case. Nevertheless, the system Green's function is still a useful quantity since it can be shown that its elements $\frac{\delta u_i(x,t)}{\delta J_j(x',t')}$ have a significant bearing on system stability behavior. The variable $J_j(x',t')$ is the perturbing flux of component j at position x' and time t'. Rather than proceed further within the complete generality of Eq. (3), it seems prudent to illustrate the relevant techniques on more specific equations. In particular, Section II below will treat the initial-boundary value type problems arising in chemical kinetics and biological transport. As a second illustration in Section III will treat sensitivity analysis in quantum mechanical scattering theory which exhibits an interesting singular perturbation type problem. Additional aspects of these problems can be found in the current literature[6].

II. Functional Sensitivity Analysis of Reaction-Diffusion Systems.

A. Elementary Sensitivities

The equations of chemical kinetics or biosystem dynamics can take a variety of forms but they generally involve both transport and kinetic flux terms. For simplicity we will consider the problem to be confined to one spatial dimension, although higher dimensional generalizations may be readily formulated. As a prototypical problem the following equations describe both reaction through R_i and diffusion

$$\frac{\partial}{\partial t} c_i = \frac{\partial}{\partial x} D_i(x,t) \frac{\partial}{\partial x} c_i + R_i(\underline{c},\underline{k};x,t), \quad i=1,2,\ldots N \tag{5}$$

where $c_i(x,t)$ is the i-th species concentration and $D_i(x,t)$ is the corresponding diffusion coefficient. All of the parameters entering into the generally non-linear rate term R_i are represented by elements of the vector $\underline{k}$. The initial and boundary conditions (assumed to be linear) for this system of equations is given by

$$c_i(x,0) = c_{oi}(x) \tag{6}$$

$$\left[A_i(x,t) \frac{\partial c_i}{\partial x} + B_i(x,t)c_i\right]_{x=x_b(j)} = c_i(x_b(j),t), \quad j=1,2 \tag{7}$$

with the system confined on the range $x_b(1) \leqslant x \leqslant x_b(2)$. The functional variation of Eq. (5) will produce the following differential equation for the sensitivity densities in analogy with Eq. (4)

$$\left[\frac{\partial}{\partial t} - \frac{\partial}{\partial x}\bar{D}_i(x,t)\frac{\partial}{\partial x}\right]\frac{\delta c_i(x,t)}{\delta\alpha_j(x',t')} - \sum_{k=1}^{N}\left(\frac{\partial\bar{R}_i}{\partial c_k}\right)\frac{\delta c_k(x,t)}{\delta\alpha_j(x',t')} = g_{ij}(x,t;x',t') \tag{8}$$

where

$$g_{ij} = \begin{cases} -\delta_{ij}\dfrac{\partial\bar{c}_i}{\partial x'}(x',t')\dfrac{\partial}{\partial x'}\delta(x-x')\delta(t-t'), & \alpha_j \equiv D_j \\[2ex] \dfrac{\partial\bar{R}_i}{\partial k_j}(x,t)\delta(x-x')\delta(t-t'), & \alpha_j \equiv k_j \\[2ex] 0 & , \alpha_j \neq D_j, \text{ or } k_j \end{cases} \tag{9}$$

The bar over the appropriate variables implies that they are evaluated with respect to the nominal or unperturbed system. In general all of the parameters are considered as elements of the vector $\underline{\alpha}$ whether they reside explicitly in the differential equation or in the external conditions of Eqs. (6) and (7). Complete specification of Eq. (8) requires variation of the initial and boundary conditions in Eqs. (6) and (7), and this can be carried out quite straightforwardly.

The numerical solution of Eq. (8) for the system sensitivities can be a complex task, but it should be noted that these equations are in fact generally easier solved than the original kinetic equations which are typically non-linear. Regardless of the numerical procedure implemented to solve Eq. (8) it is useful to consider the system Green's function defined as the solution of the following differential equations

$$[-\frac{\partial}{\partial t} - \frac{\partial}{\partial x}\bar{D}(x,t)\frac{\partial}{\partial x}]G_{ij}(x,t;x',t') - \sum_{k=1}^{N}\frac{\partial \bar{R}_i}{\partial c_k}G_{kj}(x,t;x',t')$$
$$= \delta_{ij}\delta(x-x')\delta(t-t'),\ i,j=1,\ldots N \tag{10}$$

where for simplicity the system parameters are assumed to reside in the body of the differential equation (5) and therefore we have

$$G_{ij}(x,t';x',t') = \delta_{ij}\delta(x-x') \tag{11a}$$

$$[\bar{A}_i(x,t)\frac{\partial}{\partial x}G_{ik} + \bar{B}_i(x,t)G_{ij}]\Big|_{x=\bar{x}_b(\ell)} = 0,\quad \ell = 1,2 \tag{11b}$$

Since Eq. (8) is linear, the solution may be immediately expressed as an integral over the Green's function times the inhomogeneity defined in Eq. (9). It is apparent from the structure of Eqs. (8) and (10) that elements of the Green's function matrix may be interpreted as flux response functions

$$G_{ij}(x,t;x',t') = \frac{\delta c_i(x,t)}{\delta J_j(x',t')} \tag{12}$$

with the dimensions $(\text{length})^{-1}$.

Before proceeding further, it is useful to consider a concrete example which we will take here as a set of linear reaction diffusion equations with constant coefficients

$$\frac{\partial c_1}{\partial t} = -k_1c_1 + k_2c_2 + D_1\frac{\partial^2 u_1}{\partial x^2}$$
$$\frac{\partial c_2}{\partial t} = k_1c_1 - (k_2+k_3)c_2 + k_4c_3 + D_2\frac{\partial^2 c_2}{\partial x^2} \tag{13}$$
$$\frac{\partial c_3}{\partial t} = k_3c_2 - k_4c_3 + D_3\frac{\partial^2 c_3}{\partial x^2}$$

with

$$c_i(x,0) = w_i\delta(x-q_i)$$
$$\lim_{x\to\pm\infty} c_i(x,t) = 0 \tag{14}$$

This problem can be analytically solved, and it serves as a useful illustration of the sensitivity concepts. The nominal solution vector satisfying Eqs. (13) and (14) is

$$\underline{\bar{c}}(x,t) = \frac{1}{2\sqrt{\pi \underline{\underline{D}}t}} \exp[\underline{\underline{K}}t - \frac{1}{4\underline{\underline{D}}t}(x\underline{\underline{1}} - \underline{\underline{q}})^2]\underline{w} \qquad (15)$$

where $\underline{\underline{K}}$ is the matrix of the rate coefficients in Eq. (13), the matrices $\underline{\underline{q}}$ and $\underline{\underline{D}}$ are diagonal with elements q_i and D_i, respectively, and the vector $\underline{w}$ has components defined in Eq. (14). As an illustration of the Green's function in Eq. (10) - (12), Figure 1 presents the

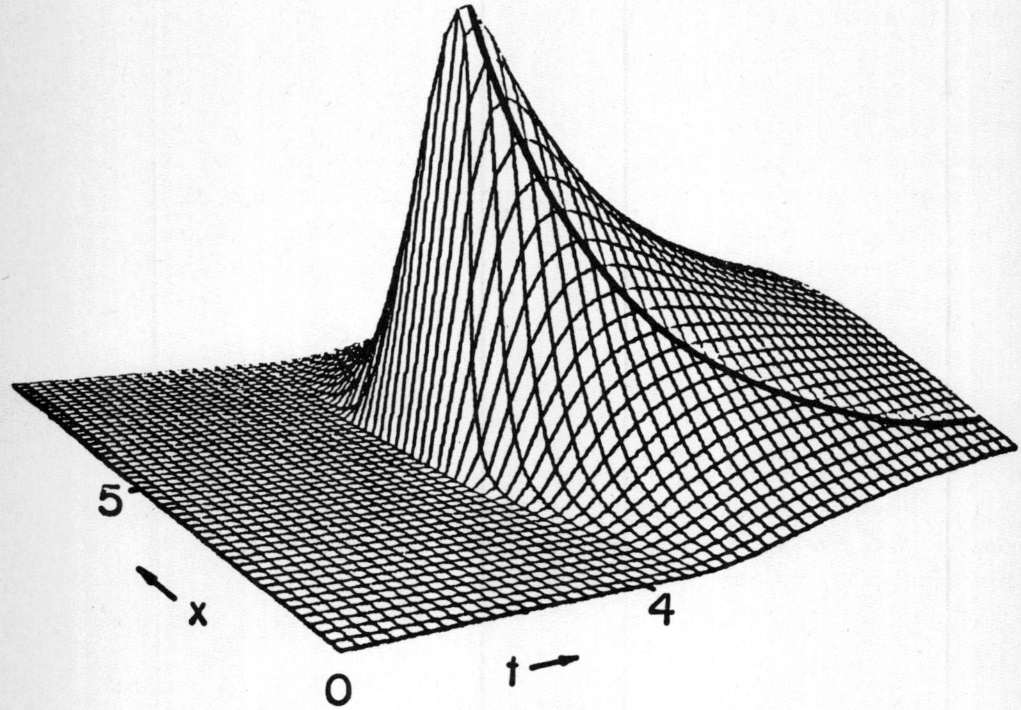

Figure 1

An element of the Green's function matrix $G_{12}(x,t;5,4)$ for the elementary sensitivity analysis of the linear reaction/diffusion problem given by Eqs. (13) and (14). Here we take $\bar{D}_1 = \bar{D}_2 = \bar{D}_3 = 1.0$, $\bar{k}_1 = 1.0$, $\bar{k}_2 = 2.0$, $\bar{k}_3 = 3.0$, $\bar{k}_4 = 4.0$, $\bar{w}_1 = \bar{w}_2 = \bar{w}_3 = 10.0$, $q_1 = q_2 = q_3 = 1.0$ and fix $x' = 5.0$, $t' = 4.0$. G_{12} reaches a maximum of 0.156 at $x = x' = 5.0$, $t = 4.2$. The solid line is the locus of maxima for cuts through the surface at constant x.

element G_{12} for the problem defined by Eq. (13)[7]. The plot clearly illustrates the point that a disturbance of $c_2(x',t')$ has its maximum effect in the neighborhood of the point x',t' for the species $c_1(x,t)$. The actual maximum point of influence occurs at $t > t'$ due to the finite time required for propagation of the disturbance through the differential equations. In addition for values of x such that $|x| >> |x'|$ it is evident that an additional delay time results for the disturbance at such distant points to be maximally felt (The locus of these points is denoted by the heavy bold line in the figure). Systems of equations exhibiting non-linear structure would be expected to show even richer response surfaces than that of Figure 1. Considering now just the simple temporal analog to Eq. (13) (i.e., the spatial derivative terms are absent) we may calculate the elementary sensitivity coefficients in a straightforward fashion. Figures 2 and 3 illustrate two cases of such sensitivity densities[7]. In both figures the densities are zero for t=t' due to causality considerations. Both figures show a certain degree of similarity with one significant difference; the density in Figure 2 peaks along the diagonal t=t' while in Figure 3 the peak occurs

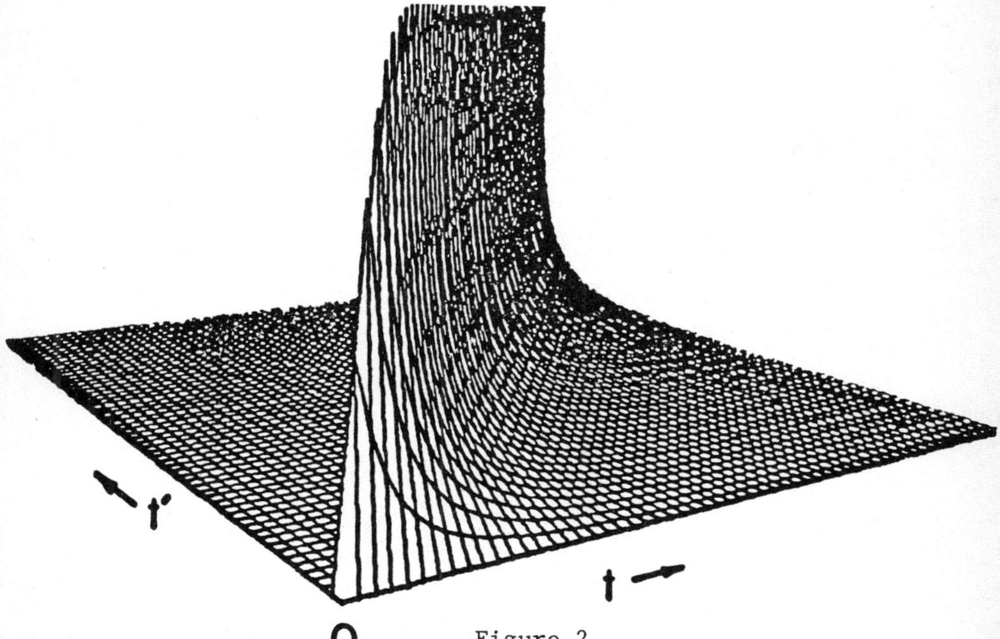

Figure 2

$\frac{\delta c_3}{\delta k_3}(t,t')$ for the linear kinetics system described in Eq. (13) without diffusion. Here we have taken the reference parameter values to be $\bar{k}_1 = \bar{k}_2 = \bar{k}_3 = \bar{k}_4 = 1$ and $c_1(t=0) = 1.0$, $c_2(t=0) = c_3(t=0) = 0.0$. The density $\frac{\delta c_3}{\delta k_3}$ asymptotically reaches a maximum of .333 (in the dimensionless units given by the reference parameters above) along the diagonal $t=t'$. For $t < t'$, $\frac{\delta c_3}{\delta k_3} = 0$ by causality; for $t >> t'$ we also have $\frac{\delta c_3}{\delta k_3} = 0$ because the system loses memory after a sufficiently long time t of the perturbation in k_3 at time t'. Expected behavior is illustrated at a delayed time $t=t'+\tau, \tau > 0$. This behavior results from the fact that the parameter k_3 enters directly into the differential equation for the third species c_3. Therefore, c_3 and k_3 are "directly" coupled and no delay time results from a perturbation in k_3. On the other hand, the parameter k_4 does not directly enter into the differential equation for c_1 and the density of Figure 3 reflects this time delay for the disturbance of k_4 at time t' to propagate through the differential equation system and produce a maximum disturbance in c_1. For sufficiently long times in both figures, the sensitivity densities take on a fixed form reflecting the fact that in that regime the densities correspond to responses of the system around its steady state equilibrium point.

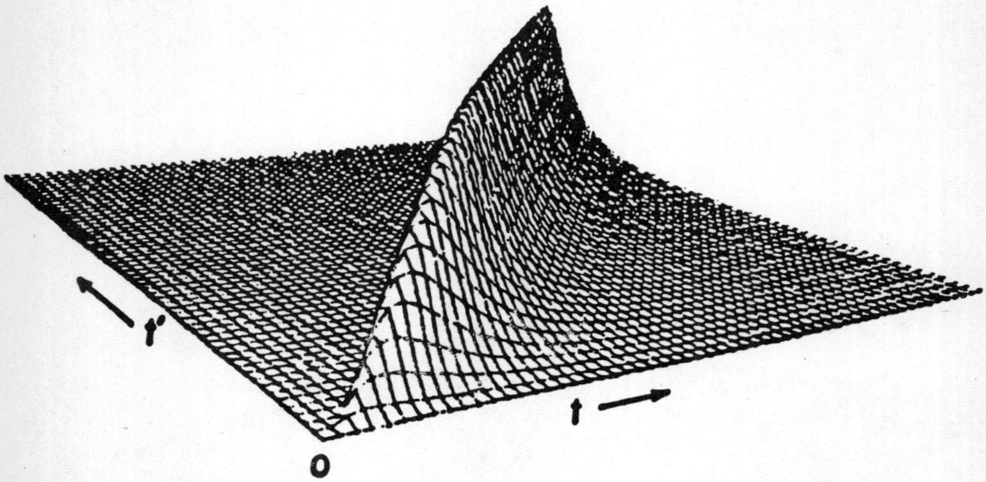

Figure 3

$\frac{\delta c_1}{\delta k_4}(t,t')$ for the same conditions given in Figure 2. The maximum plotted value of $\frac{\delta c_1}{\delta k_4}$ is 0.064 at (t = 5.6, t' = 5.4). This quantity reaches a maximum along a ridge for $t-t' \approx 0.2$. Qualitatively, Figures 2 and 3 are similar, with Figure 3 showing an important time delay between the perturbation at time t' and the observation at time t of maximum sensitivity. This time delay is due to the finite time of chemical reaction before the effect of a perturbation in k_4 can be felt by c_1.

The simple illustrations above serve to indicate the detailed information content available in sensitivity densities even for simple linear systems. Much work remains to be done on the development of computer codes for implementing the theory, and perhaps the most fruitful area for immediate pursuit may be in pure boundary value problems where there are indications that highly efficient coding can be achieved[8]. It is beyond the scope of this paper to detail explicit aspects of the numerical implementation.

B. Derived Sensitivities

By definition the sensitivity densities produced by solving Eq. (8) correspond to a dependent variable vector $\underline{c}(x)$ and an independent variable vector $\underline{\alpha}(x,t)$. This correspondence is quite natural given the original way the problem is defined in Eqs. (5) - (7). However, there is ample reason to consider the rearranged set of independent and dependent variables produced by exchanging members of the originally defined set. In this fashion we may calculate the "derived" sensitivities

$$\left(\frac{\delta\alpha_j(x,t)}{\delta c_i(x',t')}\right)^d, \quad \left(\frac{\delta\alpha_\ell(x,t)}{\delta\alpha_{\ell'}(x',t')}\right)^d, \quad \left(\frac{\delta c_i(x,t)}{\delta\alpha_j(x',t')}\right)^d \quad \text{and} \quad \left(\frac{\delta c_i(x,t)}{\delta c_j(x',t')}\right)^d$$

where the superscript d is used to indicate their derived nature. These gradients address a variety of questions corresponding to the situation where a subset of the concentrations may be assumed "measured" in exchange for a corresponding number of parameters which are now considered as dependent variables. Formally this operation may be arrived at by exchanging s dependent and independent variables where $s \leq N$ and $s \leq M$ with M being the number of original system parameters. With this change, the new dependent and independent variables are[4,7]

$$\underline{c}^d \equiv \{\alpha_1(x,t)\cdots\alpha_s(x,t), \quad c_{s+1}(x,t)\cdots c_N(x,t)\} \tag{16a}$$

$$\underline{\alpha}^d \equiv \{c_1(x,t)\cdots c_s(x,t), \ \alpha_{s+1}(x,t)\cdots\alpha_M(x,t) \tag{16b}$$

where there is no loss of generality in choosing to exchange the first s variables. It is convenient to introduce new notation for the elementary and derived sensitivity densities,

$$S_{ij}(x,t;x',t') = \frac{\delta c_i(x,t)}{\delta\alpha_j(x',t')} \tag{17a}$$

$$S^d_{ij}(x,t;x',t') = \frac{\delta c_i{}^d(x,t)}{\delta\alpha_j{}^d(x',t')} \tag{17b}$$

Using these definitions and the variables exchange stated in Eq. (16), one may readily obtain the following set of integral equations for the derived sensitivities

$$\sum_{k=1}^{s} \iint dx''dt''\, S_{ik}(x,t;x'',t'')S^{d}_{kj}(x'',t'';x',t') = \mu_{ij}(x,t;x',t') , \tag{18}$$

where

$$\mu_{ij}(x,t;x',t') = \begin{cases} -\delta_{ij}\delta(\underset{\sim}{x}-\underset{\sim}{x}')\delta(t-t'), & 1\leq i\leq s \text{ and } 1\leq j\leq s \\ -S_{ij}(x,t;x',t'), & 1\leq i\leq s \text{ and } s+1\leq j\leq M \\ S^{d}_{ij}(x,t;x',t'), & s+1\leq i\leq N \text{ and } 1\leq j\leq s \\ S^{d}_{ij}(x,t;x',t') \\ -S_{ij}(x,t;x',t'), & s+1\leq i\leq N \text{ and } s+1\leq j\leq M \end{cases} \tag{19}$$

The name "derived" follows from the fact that these new sensitivities are determined from the original elementary set through these integral equations.

The latter comment on the origin of the word derived corresponds to the historical way these coefficients were considered. A more convenient route to obtaining these derived sensitivities is directly through a set of differential equations which does not require prior knowledge of the elementary sensitivities[7]. These differential equations may be obtained by operating on Eq. (18) with the same differential operator appearing on the left hand side of Eq. (8) to produce the following matrix equation

$$\underline{\underline{L}}\ \underline{\underline{S}}^{d} = \underline{\underline{I}} , \tag{20}$$

where the parameters considered for interchange are those only explicitly residing in the original differential Eq. (5). The operator $\underline{\underline{L}}$ has the following form

$$L_{ij} = \begin{cases} f_{ij}, & 1\leq i\leq s \text{ and } 1\leq j\leq s \\ -f_{ij}, & s+1\leq i\leq N \text{ and } 1\leq j\leq s \\ \dfrac{\partial \bar{R}_i}{\partial c_j}, & 1\leq i\leq s \text{ and } s+1\leq j\leq M \\ \delta_{ij}\left[\dfrac{\partial}{\partial t} - \dfrac{\partial}{\partial x}\bar{D}_i\dfrac{\partial}{\partial x}\right] - \dfrac{\partial \bar{R}_i}{\partial c_j}, & s+1\leq i\leq N \text{ and } s+1\leq j\leq M \end{cases} \tag{21}$$

with

$$f_{ij} = \begin{cases} \dfrac{\partial \bar{R}_i}{\partial k_j}, \ \alpha_j = k_j \text{ and } i=1,\dots N \\ \delta_{ij}\left[\dfrac{\partial^2 \bar{c}_j}{\partial x^2} + \dfrac{\partial \bar{c}_j}{\partial x}\dfrac{\partial}{\partial x}\right], \ \alpha_j = D_j \text{ and } i=1,\dots s \\ 0, \ \alpha_j = D_j \text{ and } i = s+1,\dots N \end{cases} \tag{22}$$

The inhomogeneity on the right hand side of Eq. (20) has the following form

$$I_{ij} = \begin{cases} \left\{\delta_{ij}\left[\dfrac{\partial}{\partial t} - \dfrac{\partial}{\partial x}\bar{D}_i\dfrac{\partial}{\partial x}\right] - \dfrac{\partial \bar{R}_i}{\partial c_j}\right\}\delta(x-x')\delta(t-t'), & 1 \leqslant i \leqslant s \text{ and } j \leqslant j \leqslant s \\ \dfrac{\partial \bar{R}_i}{\partial c_j}\delta(x-x')\delta(t-t'), & s+1 \leqslant i \leqslant N \text{ and } 1 \leqslant j \leqslant s \\ -g_{ij}(x,t;x',t'), & 1 \leqslant i \leqslant s \text{ and } s+1 \leqslant j \leqslant M \\ g_{ij}(x,t;x',t'), & s+1 \leqslant i \leqslant N \text{ and } s+1 \leqslant j \leqslant M \end{cases} \tag{23}$$

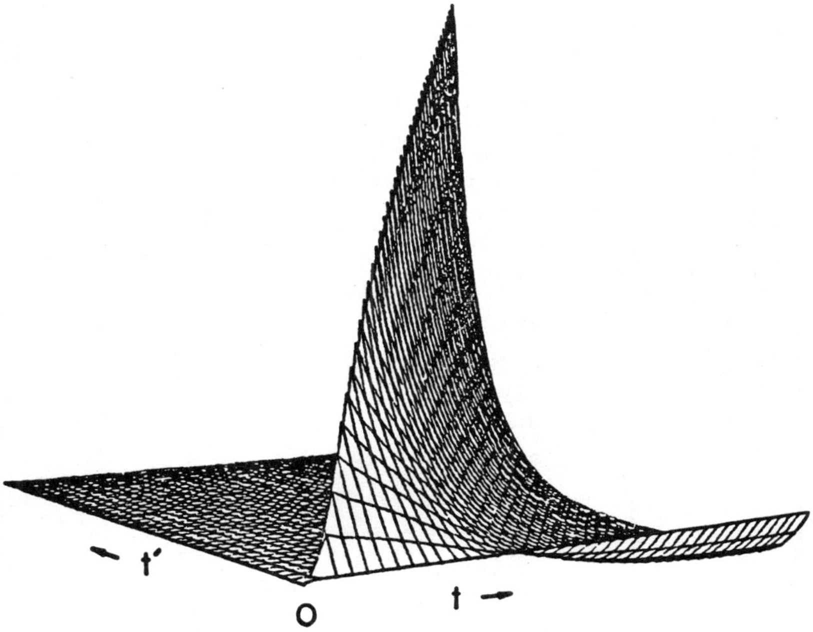

Figure 4

$\frac{\delta D_1}{\delta k_4}(t,t')^d$ for the parameter values: $\bar{k}_1 = 1.0$, $\bar{k}_2 = 2.0$, $\bar{k}_3 = 3.0$, $\bar{k}_4 = 4.0$, $w_1 = w_2 = w_3 = \bar{D} = 1.0$, $Q_1 = Q_2 = Q_3 = .52$, $x = 1.0$, $x' = 1.05$ and $0 \leqslant t \leqslant 1.0$, $0.0125 \leqslant t' \leqslant 1.0125$ with a grid spacing of 0.025. The density is singular at $t = t'$, along the diagonal and becomes small and even negative for large t, small t'.

$\underline{\underline{g}}$ is defined in Eq. (9). The system of Eqs. (20) partially decouples since it can be shown that S^d_{ij}, $1 \leq i \leq s$ and $1 \leq j \leq M$ may be explicitly expressed in terms of the remaining (lower portion) of the matrix S^d_{ij}, $s+1 \leq i \leq N$ and $1 \leq j \leq M$. Appropriate initial and boundary conditions may also be derived for Eq. (20). A Green's function approach can be applied to solve these equations since they again take on a linear form.

As a specific illustration of derived sensitivities, we may return again to the model defined in Eq. (13). The exchange of variables in Eq. (16) was achieved[7] by interchanging the diffusion coefficient D_1 and the species concentration c_1. Figure 4 shows a particular derived sensitivity density $\frac{\delta D_1(1,t)}{\delta k_4(1.05,t')}$ as a function of t and t'. It is apparent from the figure that D_1 is most closely correlated with k_4 along t=t' at which point c_1 has been "observed" to take on its nominal value $\bar{c}_1(1.05,t')$. Much further work including computations needs to be done on derived sensitivities in order to fully appreciate their meaning and usefulness.

III. Quantum Scattering Theory

The analysis presented in Section II can be carried over to a wide variety of problems having an initial-boundary value type nature. In particular, we may consider the time independent Schrodinger equation

$$[\nabla^2_{\underline{R}} - \frac{2\mu}{\hbar^2}(H_{int}(\underline{r}) + V(\underline{r},\underline{R}) - E)]\Phi^i(\underline{r},\underline{R}) = 0 \qquad (24)$$

where $\underline{R}$ is the scattering coordinate, $\underline{r}$ represents all internal non-scattering coordinates associated with the part of the Hamiltonian $H_{int}(r)$, V is the interaction potential, E is the total energy, μ is the reduced mass and Φ^i is the total wavefunction which is labelled by the incident channel index i. The problem of concern here is to calculate the scattering amplitude for making an inelastic or elastic transition amongst the other internal state ϕ_j of the system defined by

$$\begin{aligned} H_{int}\phi_j(\underline{r},\underline{\Omega}) &= \varepsilon_j\phi_j(\underline{r},\Omega) \\ \nabla^2_{\underline{\Omega}}\phi_j(\underline{r},\underline{\Omega}) &= \frac{1}{R^2}\,\ell(\ell+1)\phi_j(\underline{r},\underline{\Omega}) \end{aligned} \qquad (25)$$

where $\underline{\Omega}$ are the polar angles associated with the scattering vector coordinate $\underline{R}$, and ℓ is the orbital angular momentum quantum number.

The total wavefunction may be expanded in the set of eigenfunctions introduced in Eq. (25) to produce

$$\Phi^{i}(\underline{r},\underline{R}) = \frac{1}{R} \sum_{j} \phi_j(\underline{r},\underline{\Omega})\psi_{ji}(R) \tag{26}$$

The scattering matrix S_{ji} is defined in terms of the boundary conditions on the expansion coefficients of Eq. (26)

$$\psi_{ji}(R) \sim -\delta_{ji}k_i^{-\frac{1}{2}} \exp(-k_i R) + k_j^{-\frac{1}{2}} \exp(ik_j R)S_{ji} \tag{27}$$

where

$$k_j^2 = \frac{2\mu}{\hbar^2}(E-\varepsilon_j)$$

Substitution of Eq. (26) into Eq. (24) will lead to a set of coupled differential equations for the unknown matrix of expansion coefficients $\underline{\underline{\psi}}(R)$ and matching to the form of Eq. (27) will produce the scattering matrix $\underline{\underline{S}}$. For many practical problems this is now a standard prescription which may be numerically implemented[9]. Our purpose here is to understand how the scattering matrix may be functionally related to the Hamiltonian or portions of it[10]. In particular, we address this question by considering the functional derivative

$$\frac{\delta S_{ij}}{\delta\alpha_k(\gamma_1',\ldots,\gamma_{m_k}')} \tag{28}$$

where $\alpha_k(\gamma_1',\ldots\gamma_{m_k}')$ is the k-th system parameter entering the Hamiltonian and shown to be explicitly dependent on the subset of coordinates $\gamma_1',\ldots\gamma_{m_k}'$. In principle, we might seek system sensitivity densities by first calculating the densities for the wavefunction followed by extraction of the desired results in Eq. (28). This procedure has a singular perturbation nature if the varied parameters reside in $H_{int}(\underline{r})$ since the eigenvalues of this latter operator effect the asymptotic wavevector components. Therefore, a more direct route to calculating Eq. (28) is needed and the derivation may be implemented by considering the system Jost functions[11]. The result consists of the following sensitivity density[10]

$$\frac{\delta S_{ij}}{\delta\alpha_k(\gamma_1'\cdots\gamma_{m_k}')} = \frac{1}{2i}\left\{\sum_n \left[W_R\left[\frac{\delta f_{in}^-}{\delta\alpha_k(\gamma_1'\ldots\gamma_{m_k}')},\psi_{nj}\right]\right.\right.$$
$$\left.\left. - \sum_k \bar{S}_{ik}W_R\left[\frac{\delta f_{kn}^+}{\delta\alpha_k(\gamma_1',\ldots\gamma_{m_k}')},\psi_{nj}\right]\right]\right. \tag{29}$$
$$- \sum_{q,s}\int d\underline{r}d\underline{R}\, \bar{\psi}_{qi}(R)\bar{\phi}_q^*(\underline{r},\underline{\Omega})\, \frac{\partial\bar{G}}{\partial\alpha_k(\gamma_1',\ldots\gamma_{m_k}')}\, \bar{\phi}_s(\underline{r},\Omega)\bar{\psi}_{sj}(\Omega)\delta(\gamma_1-\gamma_1')\cdots\delta(\gamma_{m_k}-\gamma_{m_k}')$$

where

$$f^{\pm}_{kn} = \delta_{kn} k_n^{-\frac{1}{2}} \exp(\pm i k_n R)$$

$$\bar{G}(\underline{r},\underline{R}) = \frac{2\mu}{\hbar^2}[H_{int}(\underline{r}) + \bar{V}(r,\underline{R})] + \nabla^2_{\underline{\Omega}}$$

and W_R is the radial Wronskian function. The first two terms in Eq. (29) enter only when considering parameters that influence the asymptotic wave vectors.

The general result of Eq. (29) has two interesting limiting forms corresponding to varying V and H_{int} respectively. In the first case, one may show that varying V produces the following result

$$\frac{\delta S_{ij}}{\delta V(\underline{r}',\underline{R}')} = \frac{-\mu}{i\hbar^2}\left[\sum_q \bar{\psi}^+_{iq}(R')\bar{\phi}_q(\underline{r}',\underline{\Omega}')\right]^* \left[\sum_q \bar{\phi}_s(\underline{r}',\underline{\Omega}')\bar{\psi}_{sj}(R')\right] \quad (30)$$

This rather simple result is similar but generally not proportional to the system density. An analogous, but somewhat more complicated expression may be derived for $\frac{S_{ij}}{H_{int}(r)}$ where the first two terms of Eq. (29) will remain. Figure 5 illustrates Eq. (30) for a model problem consisting of the bound system described by a square well interacting with a particle by a potential having the following form

$$V(\underline{r},\underline{R}) = \begin{cases} V_{int}(\underline{r}), & R \leqslant a \\ 0, & R \leqslant a \end{cases} \quad (31)$$

This can be thought of as a square well inelastic scattering system. An interesting point seen in the figure is that the sensitivity density can change sign illustrating the point that a given S-matrix element may be increased or decreased by a complex set of potential variations occuring throughout the coordinate space. This behavior clearly illustrated the difficulty of inverting scattering information back to a uniquely defined potential.

Derived sensitivities may be considered in an exactly parallel fashion to that discussed in Section II and the detailed development will not be given here. Thusfar, no such derived calculations have been carried out for scattering, but there is an interesting question worth serious pursuit with this procedure. In the inversion of scattering information one seeks to determine the potential form scattering cross sections. and a perspective in this regard is addressed by the "inverse" of the sensitivity density arising in Eq. (30)

$$\frac{\delta V(\underline{r}',\underline{R}')}{\delta S_{ij}} \tag{32}$$

Care is needed in facilitating this calculation since exchange of the S-matrix elements and the potential function involves a discrete to continuous mapping[12]. It appears that the easiest way to achieve this transformation is by discretization or projection of the potential function onto a basis set of functions. This will produce

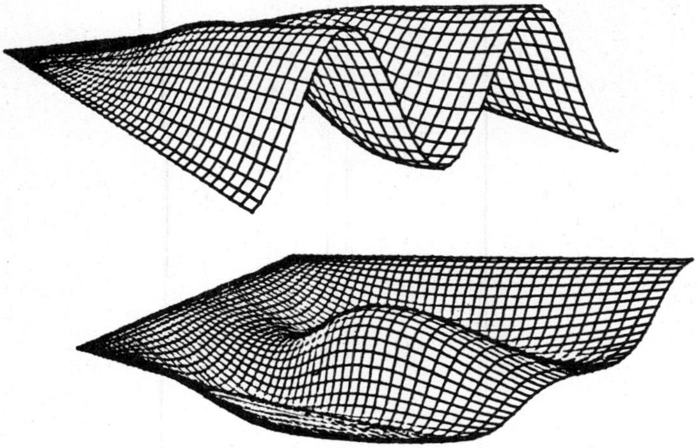

Figure 5

The real (a) and imaginary parts of the sensitivity density $\frac{\delta S_{12}}{\delta V(r,R)}$ for the model problem defined in Eq. (31). The internal coordinate r corresponding to bound motion covers the range $-\pi/2 \leqslant r \leqslant \pi/2$, and the scattering coordinate R is shown over the interaction potential region $0 \quad R \quad a = 1$. The total energy $E = 5$, the wave vector are $k_1 = 2$, $k_2 = 1$, and the potential matrix elements for $R \quad a$ are taken as $\frac{2\mu}{\hbar^2} V_{ij} = 1$.

a discrete-to-discrete mapping which should be practical to implement. Although the derived density arising in Eq. (32) does not in fact represent a scattering inversion procedure, it should nevertheless provide valuable information on how various S-matrix elements influence a portion of the potential surface.

IV. Conclusion

This paper has attempted to present an overview of certain aspects of functional sensitivity analysis with particular applications towards kinetics and quantum scattering theory. There is an important need at this point to perform practical numerical calculations in order to gain further insight in how to interpret and utilize the sensitivity densities. The generality of these sensitivity concepts and especially their ability to handle non-linear differential equations, implies that they should be a potentially valuable tool for treating a wide variety of problems besides those directly discussed in this paper.

References

1. R. Tomovic and M. Vukobratovic, General Sensitivity Theory, American Elsevier, New York, 1972.

2. V. Volterra, Theory of Functionals, Dover Publications, New York, 1959.

3. D. Cacuci, J. Math. Phys. 22, 2794 (1981).

4. M. Demiralp and H. Rabitz, J. Chem. Phys. 75, 1810 (1981).

5. M. Kramer, J. Calo and H. Rabitz, Appl. Math. Modelling, 5, 432 (1981).

6. H. Rabitz, M. Kramer and D. Dacol, Ann. Rev. Phys. Chem. 1983, in press; H. Rabitz, Computers in Chemistry, 5, 167 (1981); J. Tilden, V. Costanza, G. McRae and J. Seinfeld, in Modelling of Chemical Reaction Systems, editors K. Ebert, P. Deuflhard and W. Jager, Springer-Verlag, Berlin, 1981.

7. R. Larter, M. Kobayashi and H. Rabitz, J. Chem. Phys. 79, 692 (1983).

8. C. Irwin and T. O'Brien, DOE/METC Technical Report 82-53, Morgantown Energy Technology Center, Morgantown, West Virginia, 1982; H. Saito and L. Scriven, J. Comp. Phys. 42, 53 (1981); Y. Reuven, M. Smooke and H. Rabitz, to be published.

9. See, for example, H. Massey, Atomic and Molecular Collisions, Halstead Press, New York, 1979.

10. R. Larter and H. Rabitz, Phys. Rev. A., in press.

11. L. Eno and H. Rabitz, J. Chem. Phys. 71, 4824 (1979).

12. B. Guzman and H. Rabitz, to be published.

ACKNOWLEDGEMENT

The author would like to acknowledge support from the Office of Naval Research.

UNIQUENESS FOR GRADIENT METHODS IN ENGINEERING OPTIMIZATION

Vadim Komkov
Department of Mathematics
Winthrop College
Rock Hill, S.C. 29733

and

Carlton Irwin
Department of Mathematics
West Virginia University
and Department of Energy, METC
Morgantown, WV 26506

ABSTRACT

This paper deals with optimization of constrained functionals with distributed parameters. The functionals may be chosen for identifying parameters that optimize performance or minimize energy forms, or for computation of eigenvalues via the Rayleigh quotient. Constraints may represent design limitations, extremes of operating conditions, or state equations for dynamical systems. The intrinsic nature of iterative solution methods for functional minimization, the functional sensitivity analysis, and state function sensitivity analysis have been the subject of extensive research. Using simple examples from engineering, this paper points out some pitfalls for gradient-type computational methods particularly in connection with computing of unstable processes or eigenvalues via minimization of a constrained Rayleigh quotient. Auxiliary conditions involving energy levels of the system for constrained problems are suggested as indicators of existence of multiple gradient directions.

ACKNOWLEDGEMENT

This work was made possible by a Grant from the Department of Energy, M.E.T.C. Center, West Virginia.

The first author was supported by NSF Grant CEE 80-567.

1.0 INTRODUCTION AND FORMULATION OF THE PROBLEM

Sometime in the late-1970's, many authors studying problems of engineering optimization came to the conclusion that difficulties encountered in computational procedures could be attributed to lack of smoothness in the cost functional, or in constraints assigned to the problem. Dynamic optimization problems of elastic structures, and of optimization of structures or mechanical systems against loss of stability were studied by many authors who derived algorithmic procedures which seemed to have some inherent difficulties. Examples of papers in which these difficulties were discussed are [5], [7], and [8]. In problems involving loss of stability, bifurcation of solutions, and in vibration problems a typical state equation of a system is

$$<\alpha(\underset{\sim}{u})\, y,\eta> = \zeta<\beta(\underset{\sim}{u})y,\eta> \quad , \tag{1·1}$$

where y, η are elements of suitable Sobolëv spaces W_1, W_2, α, β linear operators mapping W_1 into W_2, and ζ is an eigenvalue which is generally a complex number.

Examples of such systems can be found in many engineering texts. Consider, for example, the vibration of an elastic beam which is freely supported at both ends.

The Euler-Bernoulli assumptions, regarding bending of beams, lead to a linear partial differential equation

$$\frac{\partial^2}{\partial x^2}\left(EI(x)\,\frac{\partial^2 y}{\partial x^2}\right) + \rho A(x)\,\frac{\partial^2 y}{\partial t^2} = q(x,t).$$

The free vibration problem for a beam is modelled by the equation

$$(EI(x)\, y_0''(x))'' - w^2 \rho A(x)\, y_0(x) = 0 \tag{1·2}$$

where $' \equiv \frac{\partial}{\partial x}$, and

E is the Young's modulus (a positive constant),

I(x) moment of inertia of the cross-sectional area about the neutral axis of bending,

A(x) - area of the cross-section,

ρ - the material density (a positive constant),

$y(x,t) = y_0(x)\, e^{i\omega t}$ is the assumed form of solution of the homogeneous equation, and

q(x,t) is the applied load per unit length.

One wishes to optimize an objective functional $\phi(y,u)$, where the design parameter u may be identified in one specific case with the cross-sectional area A(x). If we assume that I(x) is a known function of A(x), we have the state equation (1·2) reduced to the form (1·1), i.e.,

$$\alpha(A(x))\,\cdot = \frac{\partial^2}{\partial x^2}\left(I(A(x))\,\frac{\partial^2}{\partial x^2}\,\cdot\,\right)$$

$$\beta(A(x)) = \rho A(x)\ .$$

The "optimal design problem" consists of optimizing some a priori assigned cost functional $\phi(y,\underset{\sim}{u})$. The state equation (1·1) is regarded as a constraint assigned to the problem. Additional constraints (such as constraints on the highest value of stress, largest permissible deflection, largest available size of a structural element, etc.) are also assigned to the optimization problem. Constraints imposed on admissible values of the design vector $\underset{\sim}{u}$ are generally dictated by manufacturing or technological limitations.

The Rayleigh quotient for the one-dimensional case is of the form

$$\omega^2 = \langle K_1 E A^a \alpha y, \alpha y(\underset{\sim}{x})\rangle_\Omega \,/\, \langle (K_2 \rho A^b - q(\underset{\sim}{x}))\, \beta y(\underset{\sim}{x}), \beta^* y(\underset{\sim}{x})\rangle_\Omega + (Q_i, \phi(y(\underset{\sim}{x}_i))_{\partial\Omega} ,$$

where the product $\langle\ ,\ \rangle_\Omega$ is the usual $L_2(\Omega)$ inner product

$$(\underset{\sim}{f}, \underset{\sim}{g}) = \int \sum_{i=1}^{n} f_i(\underset{\sim}{x}) \cdot g_i(\underset{\sim}{x})\, dx,$$

and $A(x)$ is the area of cross-section regarded as the design parameter. (See Olhoff [5]).

In the buckling problem, the proper functional analytic setting is

$$W = H_0^2\,[0,1]\,,\ (\Omega = [0,1]);\ K_1 = \tfrac{1}{2},\ K_2 = \tfrac{1}{2},\ \alpha \equiv \frac{d^2}{dx^2} = \alpha^*,$$

$\beta = \frac{d}{dx}$, $\beta^* = -\frac{d}{dx}$ for static problems and $\beta = \lambda$ (constant) for vibration problems. ϕ is a continuous function of y on $\partial\Omega$.

A geometric constraint of the type $\int_0^1 A(x)dx = 1$ is imposed.

Physically this constraint implies that the total weight is constant.
Many papers were written in the 1970's establishing necessary conditions for this problem, and for the dual problem. See, for example, [6], [10], [11], [12], and [13].

In most of those papers a simple Frechet differentiation technique produced the sensitivity formulas for optimization of design with respect to lowest eigenvalue and necessary conditions for optimality. (See also [5], [22], and [23] for modifications of this approach.)

However, implementations of these results in the form of gradient or gradient projection algorithms offered designs which were not acceptable to the design engineers. This was resolved by an observation regarding loss of differentiability which necessarily occurs in the implementation of numerical algorithms of the gradient type. We should note that certain direct techniques of W. Prager, John Taylor, and of some Prager collaborators did not suffer from this defect.

Observations concerning non-differentiability can be claimed by several authors (E. Masur [17], N. Olhoff [21], and Haug and Roussellet [19]). To illustrate the difficulties which arise in typical optimization problems, we offer a fairly simple example of optimization for a vibrating mechanical system.

While this example is very simple, it does contain all ingredients that illustrate the serious and deep difficulties associated with the occurrence of multiple eigenvalues in either discrete systems or in systems with distributed parameters.

In the specific mechanical example offered here, the reasons for these difficulties are physically obvious. Such physical reasoning is very difficult in more complex mechanical or structural systems.

For these reasons the example offered here is enlightening.

In [32] Choi and Haug offered a simple example of an optimization for a 2-degrees of freedom system shown on figure 1.

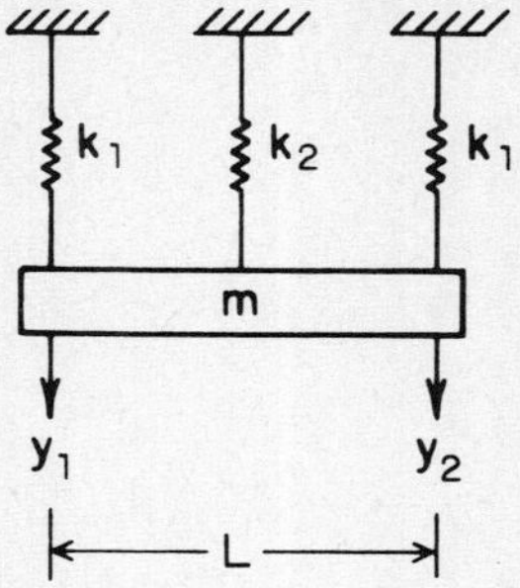

Figure 1

Two-degree of freedom spring-mass system.

We believe that our example illustrates more clearly the phenomena that occur as merging of eigenvalues interferes with the gradient algorithm.

2.0 An Example

Let us consider vibrations of a two-degrees of freedom system consisting of two elastic supports, a rigid frame, and a mass as shown on Figure 2. The rigid frame has a constant moment of inertia equal to $\rho^2 m_2$.

We assign a constraint to values of K_1, K_2, such as

$$\left(\frac{K_1 K_2}{K_1 + K_2} \right) = \text{constant.}$$

$\left(\frac{K_1 K_2}{K_1 + K_2} \right.$ is the equivalent spring constant for two parallel springs.)

The Euler-Lagrange equations of motion for this system are

$$\frac{\partial (T-V)}{\partial \underset{\sim}{q}} - \frac{d}{dt} \left(\frac{\partial T}{\partial \dot{\underset{\sim}{q}}} \right) = 0 \quad .$$

Example:

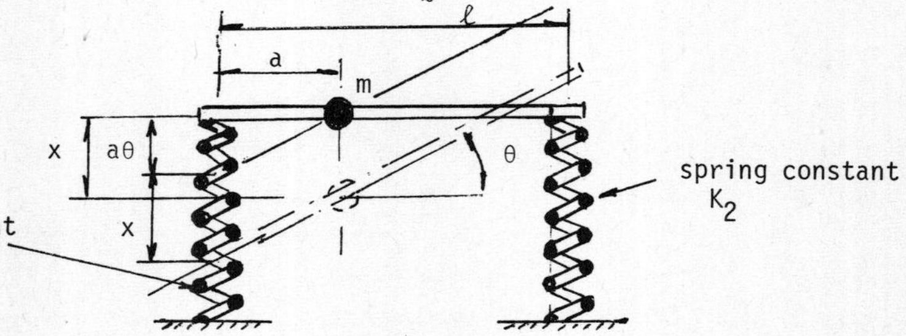

FIGURE 2

m_1 = a point mass (located at position a)
m_2 = mass of bar (uniformily distributed)
ρ = radius of inertia for the bar
$x(t)$ = vertical displacement of m_1 at time t
$\theta(t)$ = angular displacement of bar at time t
$V(t)$ = potential energy at time t = $\frac{1}{2}[K_1(x(t) + a\theta(t))^2 + K_2(x(t) - (\ell-a)\theta(t))^2]$
$T(t)$ = kinetic energy at time t
$= \frac{m_1 (\dot{x}(t))^2}{2} + \frac{m_2\rho^2}{2}(\dot{\theta}(t))^2$
K_i = spring constants
ℓ = the length of the bar
a = distance at which mass m_1 is positioned,

where $\underset{\sim}{q} = \begin{bmatrix}\theta \\ x\end{bmatrix}$, the design vector is $\begin{bmatrix} m_1 \\ m_2 \\ K_1 \\ K_2 \\ a \end{bmatrix} = \underset{\sim}{\alpha}$; we denote $\omega = \dot{\theta}$.

For a fixed design the kinetic energy is given by

$$T = \frac{m_1(\dot{x})^2}{2} + \frac{m_2\, \rho^2\omega^2}{2} , \qquad (2\cdot2^{\underline{a}})$$

and the potential energy by

$$V = \frac{1}{2}[K_1(x+a\theta)^2 + K_2(x - (\ell-a)\theta)^2] . \qquad (2\cdot2^{\underline{b}})$$

If we assume a harmonic motion, the corresponding eigenvalue problem reduces to solving the algebraic equation

$$\lambda^4 - \left(\frac{K_1 + K_2}{m_1} + \frac{K_1a^2 + K_2(\ell-a)^2}{m_2\rho^2}\right)\lambda^2 + \frac{K_1K_2\,\ell^2}{m_1m_2\rho^2} = 0 . \qquad (2\cdot3)$$

Solving (2·3) for λ we obtain

$$\lambda^2_{1,2} = \frac{B}{2} \pm \frac{\sqrt{B^2 - 4C}}{2} , \ (\lambda_1 < \lambda_2),$$

where B is the quantity inside the brackets in (2·3) and

$$C = \frac{K_1K_2\,\ell^2}{m_2\rho^2} , \qquad B^2 > 4C .$$

For the sake of simplicity, let us assume $m_1 = m_2$, i.e., the center of mass of the frame coincides with the position of the discrete mass m (as shown on Figure 2).

If, in addition, we oversimplify the problem by assumming $\rho = \sqrt{a(\ell-a)}$ (which may be in some physical cases, a very sensible assumption), we derive the following eigenvalues

$$\lambda_1^2 = \frac{K_1\ell}{m(\ell-a)} \quad , \quad \lambda_2^2 = \frac{K_2\ell}{ma} \quad . \tag{2.4}$$

The ratio of θ_{max}/x_{max} becomes $\frac{\ell}{(\ell-a)}$ in the first mode and $\frac{\ell}{a}$ in the second mode, corresponding, respectively, to λ_1^2 and λ_2^2 .

The dependence of λ_1^2 and λ_2^2 on the length a is smooth and we have

$$\frac{d\lambda_2^2}{da} = - \frac{K_1\ell}{m(\ell-a)^2} \quad ,$$

$$\frac{d(\lambda_1)^2}{da} = + \frac{K_2\ell}{ma^2} \quad .$$

The sensitivity of λ_1, λ_2 with respect to the design vector $\begin{bmatrix} m \\ K_1 \\ K_2 \\ a \end{bmatrix}$ is easily established.

It appears that subject to the constraint $\frac{K_1K_2}{K_1 + K_2}$ = constant, all we have to do in order to improve the value of λ_1 is to compute the derivative of either the eigenvalue of λ_1 or λ_1^2 with respect to the design vector while obeying the constraint specifically; if we keep values K_1 and K_2 constant, $K_1 = K_2$, and only adjust the length a, λ_1^2 increases. If we start with a value of $a < \ell/2$, we improve λ_1^2 (here $\lambda_1^2 < \lambda_2^2$) by increasing a. When $a = \frac{\ell}{2}$, (with $K_1 = K_2$) we have $\lambda_1 = \lambda_2$. Two possible values of the derivative can be chosen in the gradient algorithm. Trouble arises if we choose either one. Trying to improve the smaller eigenvalue is impossible (see figure). As we increase λ_1, we decrease λ_2, and vice versa. The problem of maximizing the lowest eigenvalue is not solvable by gradient method if we do not constantly check which mode of vibration corresponds to the smaller eigenvalue. λ_{min} is not a differentiable function of the parameter a at the value $a = \frac{\ell}{2}$ which corresponds to coalescence of the two eigenvalues.

This is a very simple example corresponding to a quite general theoretical phenomenon. Hence, curiosity may be aroused by the behavior of this system when $\lambda_1 = \lambda_2$.

The Euler-Lagrange equations, written out in full are

$$m_1\ddot{x} + (K_1 + K_2)x + (K_1 a - K_2(\ell-a))\theta = 0$$

(2.5)

$$m_2\rho^2\ddot{\theta} + (K_1 a - K_2(\ell-a))x + (K_1 a^2 + K_2(\ell-a)^2)\ \theta = 0,$$

i.e., the equations of motion reflect the coupling between the linear motion of the mass m_1 and the angular motion of the frame having the moment of inertia $I = m_2\rho^2$. They cannot be uncoupled, except for some specific choices of K_1, K_2, a, m_1, m_2.

Denoting

$$x = \hat{x}\, e^{i\omega t}$$
$$\theta = \hat{\theta}\, e^{i\omega t} \quad ,$$

we find out that(when $\theta = \hat{\theta} e^{i\omega t}$) the decoupling takes place if the corresponding eigenvalue equation has equal roots.

To clarify our analysis let us rewrite the problem in some detail.

We have a simple harmonic motion

$$x(t) = x_o\ e^{i\omega t}$$
$$\theta(t) = \theta_o\ e^{i\omega t}$$

Let us denote

$$V(x_o,\theta_o) = \max V(t) = \frac{K_1}{2}(x_o + a\theta_o)^2 + \frac{K_2}{2}\ (x_o - (\ell-a)\theta_o)^2$$

and

$$T(x)_o,\ \theta_o) = \max T(t) = \omega^2\{\frac{m_1}{2}\ x_o^2 + \frac{m_2\rho^2}{2}\ \theta_o^2\ \}$$

We identify λ with ω^2

$$\lambda_{min} = \min_{(x_o,\theta_o)} \Lambda(x_o,\theta_o) \quad ,$$

where

(2.6) $$\Lambda(x_o,\theta_o) = \frac{V(x_o,\theta_o)}{T(x_o,\theta_o)}$$

$(x_o,\ \theta_o)$ is the amplitude vector for the harmonic motion that solves the system (2.5).

Λ is the corresponding Rayleigh quotient. In our example, the Rayleigh quotient formula (2.6) is specifically given by the expression:

$$\Lambda = \frac{(x_o\ \theta_o)\begin{bmatrix} K_1 + K_2 & aK_1 - (\ell-a)K_2 \\ aK_1 - (1-a)K_2 & a^2K_1 + (\ell-a)^2K_2 \end{bmatrix}\begin{pmatrix} x_o \\ \theta_o \end{pmatrix}}{(x_o\ \theta_o)\begin{bmatrix} m_1 & 0 \\ 0 & m_2\rho^2 \end{bmatrix}\begin{pmatrix} x_o \\ \theta_o \end{pmatrix}} \qquad (2.7)$$

i.e. $\lambda = \dfrac{< y^T M y >}{< y^T N y >}$.

$$\lambda_{min} = \min_{x_o, \theta_o} \left\{ \frac{y^T M y}{y^T N y} \right\}$$

= minimal eigenvalue for $My = \lambda Ny$ in buckling problems,

= minimal eigenvalue for $Ty = \lambda y$ in vibration problems,

where

$$T = \begin{bmatrix} \dfrac{K_1 + K_2}{m_1} & \dfrac{aK_1 - (\ell-a)K_2}{m_1} \\ \dfrac{aK_1 - (\ell-a)K_2}{m_2\rho^2} & \dfrac{a^2K_1 + (\ell-a)^2K_2}{m_2\rho^2} \end{bmatrix}$$

$$Y = \begin{pmatrix} x_o \\ \theta_o \end{pmatrix} .$$

Characteristic equation for the matrix operator T is

$$0 = \lambda^2 - \left[\frac{K_1 + K_2}{m_1} + \frac{aK_1 + (\ell-a)^2K_2}{m_2\rho^2}\right]\lambda + \left(\frac{K_1 + K_2}{m_1}\right)\left(\frac{a^2K_1 + (\ell-a)^2K_2}{m_2\rho^2}\right) - \frac{[aK_1 - (\ell-a)K_2]^2}{m_1 m_2 \rho^2}$$

Again, to simplify calculations let

$m_1 = m_2 = m$

$K_1 = K_2 = K$

$\rho = \sqrt{a(\ell-a)}$,

deriving

$\lambda_1 = \dfrac{K\ell}{m(\ell-a)}$,

$\lambda_2 = \dfrac{K\ell}{ma}$

Design problem: Hold K, ℓ, m constant and determine a so as to find:

$$\Lambda = \max_a(\min(\Lambda_1, \Lambda_2).$$

That is, our design is optimal if we maximize Λ over the admissible values of a.

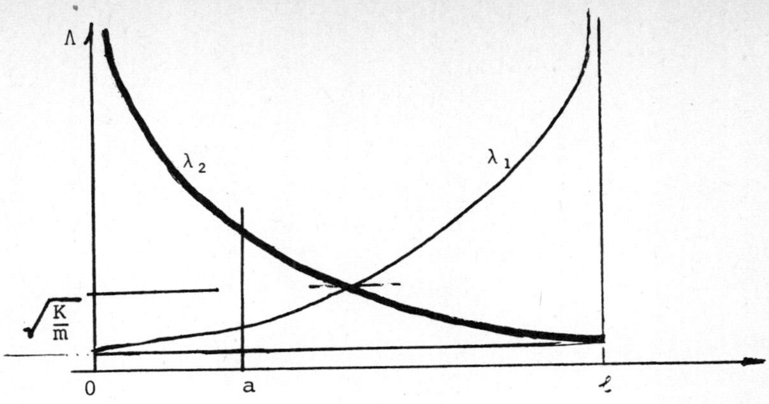

Figure 3

λ_1 and λ_2 as functions of the parameter a.

$\Lambda = \lambda_{min}\,(a) = \min\,\{\lambda_1(a),\ \lambda_2(a)\}$ is not a differentiable function of a at the point $a = \frac{\ell}{2}$.

The maximum of λ_{min} is equal in this case $K_1/(2m)$. Obviously, a similar result would be obtained (with somewhat different values) for the general case when $K_1 \neq K_2$, $m_1 \neq m_2$.

The equation (13) has equal roots when Λ is not a differentiable function of the parameters. Moreover, the eigenvalue equation decomposes into two algebraic equations, such as

$$\hat{x}\ (-m_1\lambda_1^2 + (k_1 + k_2)) = 0$$

$$\hat{\theta}\ (-m_2\rho^2\lambda_2^2 + k_1 a^2 + k_2(\ell - a)^2) = 0,$$

where it is clear that the values of λ_1 and λ_2 are completely independent of each other. For example, changing m_1, does not affect λ_2 and varying $m_2\rho^2$ does not affect λ_1. Also, λ_1 is completely insensitive to changes in the total length ℓ. L. I. Mandelstam points out in his monograph "Lectures on the theory of vibrations" that at the coalescing of eigenvalues the phases of different modes do not have to be the same. This remark is best understood if we realize that the equations of motion become decoupled.

It is interesting to note that the decoupling occurs exactly at the point of non-differentiability of the functional $\Lambda = \lambda_{min}$. Also, our remark points out some subtle errors frequently made by many engineers in the analysis of structural, mechanical or electrical resonances. Arguments asserting the existence of "secular terms", (for example, $te^{i\theta t}$ for a harmonic oscillator whenever multiple eigenvalues occur) in mechanical vibration problems go back to the monograph "Analytical Mechanics" or J. Lagrange, who made such an erroneous conclusion in both the first edition (of 1788) and in the second edition (of 1812) of that monograph despite the fact that such assertion violates the basic physical law of conservation of energy.

We take another look at the system of two Euler-Lagrange equations of motion for our mechanical system:

$$m_1 \ddot{x} + (K_1 + K_2)x + (K_1 a - K_2(\ell-a))\theta = 0,$$

$$m2\rho^2 \ddot{\theta} + (K_1 a - K_2(\ell-a))x + (K_1 a^2 + K_2(\ell-a)^2)\theta = 0.$$

Obviously, the system decouples whenever

$$K_1 a - K_2(\ell-a) = 0,$$

that is when $a = \dfrac{K_2 \ell}{K_1 + K_2} = \hat{a}$,

and for $K_1 = K_2 = K$ when $a = \frac{\ell}{2}$.

The state of the system is given by

$$x = x_0 \exp \{i(\frac{2K}{m_1})^{\frac{1}{2}} t\}$$

$$\theta = \theta_0 \exp \{i(\frac{a^2 + (\ell-a)^2 K}{m_2\rho^2})^{\frac{1}{2}} \cdot t\}$$

$$\theta = \theta_0 \exp \{ i \sqrt{\frac{a^2 + (\ell-a)^2}{m_2\rho^2}} K \cdot t\}$$

In general,

$$\lambda_{min}(a) = \min_{\underset{\sim}{u}} \Lambda(\underset{\sim}{a}([\underset{\sim}{u}])) ,$$

$$\underset{\sim}{a} = (a_1 \; . \; . \; . \; a_m) ,$$

$$\underset{\sim}{u} = (u_1 \; . \; . \; . \; u_n) .$$

At $a = \hat{a}$ we observe the following phenomena:

- Coalescence of eigenvalues

$$\lambda_1(a) = \lambda_2(a) \text{ at } a = \hat{a}$$

- Quasi-differentiability of λ_{min}, but lack of Fréchet differentiability of

  $\Lambda(\hat{a}) = \min\{\lambda_1, (a), \ldots, \lambda_n(a)\}$ at $a = \hat{a}$.
- Decoupling of the equations of motion

  at $a = \hat{a}$.
- Maximization of $\lambda_{min} = \Lambda$

  at $a = \hat{a}$.

As we have commented, the implications of our example are valid in much more complex design problems. A simple non-existence theorem for the simplest multi-dimensional case (optimization of a membrane) is given in [20]. It also illustrates the basic difficulties encountered in straight-forward implementation of necessary conditions for local optimality or in application of corresponding numerical algorithms.

The term "non-existence" does not imply non-existence of optimal design which occurs here at the point $a = \hat{a}$. It refers to the non-existence of a Fréchet derivative at the point $a = \hat{a}$ corresponding to optimal design.

Other phenomena studied by S. Antman that have been associated with bifurcation phenomena could be perhaps restated as conditions of non-differentiability at the point of optimal design. This aspect of our sensitivity study lies outside the scope of this work.

Certain aspects of these implications require a group-theoretic approach. That also lies outside the scope of our discussion.

3.0 Continuity and Differentiability

We consider variational problem

$$\langle A(u)Z, V\rangle_{L_2(\Omega)} = \zeta\langle B<(u)Z, V\rangle_{L_2(\Omega)} \qquad (3\cdot1)$$

A is a positive, strongly elliptic, linear operator from a Sobolev space H into $L_2(\Omega)$. B is also a linear operator mapping $H_2(\Omega)$ into $L_2(\Omega)$, where H_2 is a Sobolev space which may be identified with $L_2(\Omega)$.

For the class of operators considered in structural and mechanical problems discussed in this paper, it is well known that solutions are continuous functions of the parameters if one does not consider a neighborhood of a bifurcation point. In fact, much stronger results are available if one considers the location of the discrete spectrum for the beam or plate operator.

In [19], Haug and Rousselet prove that the eigenvalues are Gateaux differentiable functions of the design vector in linear problems of structural analysis.

(Comment: Continuity of the solution vector $y \in H$ with respect to the design vector

$\underset{\sim}{u}$ is understood in the accepted sense, i.e., small changes in the norm of $\underset{\sim}{u}$ produce only small changes in the norm of y. The beam and plate operators are the usual linear operators of the Euler-Bernoulli and Lagrange-Germaine theories, respectively.) However, near bifurcation point, or a point of coalescing of eigenvalues, the eigenvalues fail to be Fréchet differentiable functions of the design vector. As our simple example shows, a number of different criteria could be used to predict non-differentiability of the eigenvalues. One could study the dimension of the null space in the manner of [18], or other properties (accretive property) of the operator, or approach the mathematical formulation by considering the physics of the problem in the style of W. Prager. By improving sufficient constraints on the problem, one could minimize the danger of coalescing of eigenvalues and continue to employ direct optimization techniques.

Some engineers who observed the undersirable behavior of higher eigenvalues proposed some safeguards which would prevent "the crossover" phenomenon, or at least would prevent the physically unexplainable singularities which arose in the iterative optimization schemes.

In a series of papers N. Olhoff, J. Taylor, H. Rasmussen, E. Masur, E.J. Haug, and his callaborators established a number of algorithmic approaches designed to avoid the coalescing of eigenvalues in eigenvalue optimization problems. A simple argument offered in [20] indicates that in unconstrained problems multiple eigenvalues must arise as one gradually improves the design and that optimality criteria arrived by gradient-type procedures must lead to lack of differentiability associated with the "crossover" phenomenon.

Perhaps the simplest formal approach to the problem of preventing the crossover of multiple eigenvalue solutions is suggested in 1977, in a paper if N. Olhoff and S. H. Rasmussen [8], who introduce a modification of the Rayleigh quotient.

Again, one discusses the one-dimensional case, that is the buckling of a column. Consider the functional λ for the bimodal design S(x)

$$\lambda = \int_0^1 S^2(x)(y''_1)^2\,dx = \int_0^1 S^2(x)(y''_2)^2 dx,$$

where $y_1(x)$ and $y_2(x)$ are the first and second fundamental modes. The problem is normalized by requiring

$$\int (y'_i)^2 dx = 1,\ i = 1, 2 \quad ,$$

and

$$\int_0^1 S(x)dx = 1 \quad .$$

The variational functional is

$$\lambda^* = \int_0^1 S^2 (y_1'')^2 - \gamma_1 \int_0^1 S^2 [(y_1'')^2 - (y_2'')^2] dx$$

$$- \gamma_2 \int_0^1 ((y_1' - 1)\, dx - \gamma_3 \int_0^1 ((y_2')^2 - 1) dx$$

$$- \gamma_4 \left(\int_0^1 S dx - 1\right) \quad ,$$

where $\gamma_1, \gamma_2, \gamma_3, \gamma_4$ are Lagrangian multipliers.

Necessary conditions for a bimodal minimum are derived by Olhoff and Rasmussen in [8] in the form of a nonlinear integro-differential equation.
Their algorithm has an attractive feature. It also handles a single eigenvalue problem, i.e., the case where the buckling load is optimal while it corresponds to a single mode. This can be achieved by constraining the problem and allowing $S(x)$ to take values only in a certain range $S(x) \geq S_0 > 0$ and perhaps $S(x) \leq S_1$. A suitable choice of S_0 (and S_1) prevents the onset of the "crossover" phenomenon. Of course, one has to modify the functional λ^* by incorporating a term containing a slack variable $g^2(x) = S(x) - S_0$(see [8]). We shall not follow this line of reasoning.

In this paper we point out other possibilities of discovering non-differentiability by observing a change of sign of a certain energy term, indicating a bifurcation of solutions or a change in the dimension of the null space.
In the one-dimensional case, Komkov pointed out in [24], that branching of eigenfunctions of the beam operator can be predicted from the behavior of an energy form which is a functional of the domain. In the beam buckling case, the domain consists of the interval [0, 1]. The optimization problem discussed in [24] consisted in finding cross-sectional area $A(x)$ and the corresponding moment of inertia $I(x)$ of a beam, that maximized the first (i.e., lowest) eigenvalue. Fréchet differentiability generally does not exist and simple gradient-type algorithm could not be used in this case. Quasi-differentiability could be invoked to prove the existence of an optimal design. We mention that quasi-differentiability is defined in the Appendix 1.
For a detailed analysis of quasi-differentiability of eigenvalues in structural analysis see [25] . The problem of interest to us is the failure of the iterative gradient projection method applied to the lower eigenvalue.
If one realizes the reason for the lack of differentiability of the lowest eigenvalue, or similar reason for the lack of differentiability of an arbitrary functional mapping some subset of a Sobolev space W into the reals, one can immediately modify the problem, enlarge the set of "candidates" for eigenfunctions corresponding to one of the eigenvalues, and after making certain that this class of functions is a convex subset of W, establish the existence of subgradients and of supporting hyperplanes. Large literature exists concerning such problems of minimizing, or maximizing nondifferen-

tiable functionals. See, for example, [16] and [26].

4.0 Energy Arguments

Let us now examine the dependence of the eignevalues on the domain, i.e., in the one-dimensional case on the length and in two-dimensional case on the shape of the region. Two basic energy forms determine the existence of nontrivial solutions and the appearance of bifurcation points.

We formulate our problem in the setting of H_0^2 [0,1] Sobolev space. The functions I(x), A(x) are regarded as L_2 [o,1] functions, even if on physical grounds, only piecewise continuous functions can be regarded as admissible. The strain energy is given by

$$V = \frac{1}{2} \int_0^1 E\, I(s(x))(w''(\underset{\sim}{s}))^2\, dx = \langle Aw,w \rangle$$

where as before, $\langle\ ,\ \rangle$ denotes the inner product in the Sobolev space H_0^2 [0,1], A denotes the symmetric operator

$A = \frac{d^2}{dx^2}(EI \frac{d^2\cdot}{dx^2})$ and

where $\underset{\sim}{s}$ is the design vector, $\underset{\sim}{s} = \underset{\sim}{s}(x)$.

The work done by the axial load P is given by

$$W = \frac{1}{2} \int_0^1 P(w')^2\, dx = P\langle Bw,w \rangle_{H_0^2[0,1]}$$

if only the bending action is considered.

Let ρ be any admissible ($C^1[0,1]$) function.

The eigenvalues μ_i could be identified with energy levels

$$\mu_i \int_0^1 (\rho'(x))^2 dx$$

within the framework of the linear theory. Rayleigh's theorem states that

$$\mu_k \int_0^1 (\rho')^2 ds \geq \int_0^1 E\, I(x)(\rho'')^2 ds \quad , \qquad (4\cdot2)$$

if

$$\langle \rho, \psi_i \rangle = 0 \qquad i = 1, 2, \ldots, k-1,$$

where ψ_i are the eigenfunctions corresponding to the lowest possible energy levels, for the potential energy integral given by the usual quadratic form:

$$\int_0^1 E\, I(x)(\rho'')^2 ds$$

for admissible displacement functions ψ_i,

$$i = 1, 1, \ldots , k-1.$$

Returning to the problem of optimal design, for example, of maximizing $\mu_1(\underset{\sim}{s})$ subject to constraints

$$\int_0^1 A(\underset{\sim}{s})(x))\, dx = 1 \quad , \qquad g_i(\underset{\sim}{s}) \le 0 \quad , \quad i = 1, 2, \ldots m \quad , \tag{4·3}$$

one can start an algorithmic procedure of finding the first eigenvalue μ_1 for a given design $\underset{\sim}{s}$ and taking the gradient of the functional

$$\phi(\underset{\sim}{s}, \mu, \rho(x)) =$$

$$E \int_0^1 I(\underset{\sim}{s}(x))(\rho'')^2 dx - \mu \int(\rho')^2 dx = V(\underset{\sim}{s}, \rho) - W(\mu, \rho) \quad .$$

Bifurcation occurs if there exists a nonzero deflection function $\psi(x) \in H^2_o[0,1]$, such that $V(\psi) - W(\psi) \ge 0$. We comment that $\psi \in H^2_o\,[0,1]$ implies that $\psi(x) \in C^1[0,1]$ because of Sobolev's imbedding lemma.

Hence, if we can find a $C^1[0,1]$ trial function ψ satisfying this inequality, supercritical condition exists.

We note that $V(\psi(\underset{\sim}{s}), \underset{\sim}{s})$, $W(\psi(\underset{\sim}{s}))$ are design dependent functionals. The usual eigenvalue optimization problem consists of finding a design vector $\underset{\sim}{s}$ such that the smallest eigenvalue P(that is the critical load) is maximized, while the volume of the column is constrained, i.e.,

$$\int_0^1 A(s)\, dx = \text{constant, for the one-dimensional case.}$$

We introduce the space of admissible deflection function $\underset{\sim}{s}$, which in the one-dimensional case is $C^1[0,1]$, with second derivative square integrable on [0,1], i. e., $H^2_o[0,1]$.

The usual approach to this problem suggested in many engineering papers (see, for example, N. Olhoff [5] , for Haug and Arrora [10] improves the design by increasing iteratively the value of the Rayleigh quotient that is regarded as a function of the designs, using gradient projection method, while at the same time minimizing it over the space of admissible deflection functions. As our example has indicated, this approach collapses when improvement in the value of the lowest critical load (i.e., the fundamental eigenvalue) produces non-differentiability of the Rayleigh quotient.

One could study, similtaneously, the improvements in the trial functions approximating the eigenfunctions corresponding to the first and second eigenvalue and sound an alarm which would stop the iteration if the difference between these trial eigenvalues becomes "small." That is, we could take this difference and compare it with magnitude of the lower of the two eigenvalues. If this ratio is smaller than say .1, or some other a priori set number, the iterative algorithm is discontinued.
Alternately, one could minimize the Rayleigh quotient plus the difference between its current value and the value of the same quotient in the hyperplane orthogonal to the corresponding trial function.
An algorithmic approach based on formal computation, which has that flavor, was proposed in the Olhoff, Rasmussen paper [8].
While practical results obtained by this numerical approach looked good (see [8]), there are so many basic criticisms which could be voiced, that probably the best policy is to let the readers suggest some obvious or subtle shortcomings. But the most urgent criticism should concern the inability to pinpoint the exact place in the algorithmic sequence when "an improvement" in design is no longer an improvement. This could be obtained numerically. In fact, an obvious scheme can be suggested, based on the ideas in [8]. However, a scheme that would carry out the optimization of design while computing both the first and second corresponding eigenvalue would require large computing time and carries many seeds of error, simply, because it is so complex.
Let us, instead, propose an energy oriented approach which offers an alternative numerical check on non-differentiability of a functional. We shall concentrate on computation of eigenvalues and improvements in the magnitude of the first eigenvalue caused by iterative treatment of the design variables. It is always the best policy to start with the simplest example.
We consider the buckling problem for a column and the corresponding Rayleigh quotient

$$\mu_1 = \inf \left\{ \int_0^1 I(x)(\psi')^2 dx \Big/ \int_0^1 \psi^2 dx \right\} , \tag{4·1}$$

with orthogonality conditions

$$\int_0^1 \left(I(x)\, \psi_k(x)\, \psi_i(x) \right) dx = 0 \quad , \text{ for all } i < K \quad ,$$

where ψ_i (x) are the eigenfunctions of the eigenvalue problem

$$E(I(x)\, \psi_i')' = \mu_i \psi_1 \quad .$$

A geometric interpretation identifies the function $\psi(x) = w'(x)$ as the slope of the column.

The boundary conditions are: $\psi(0) = \psi(1) = 0$.
For an arbitrary test function $\rho(x)$ such that $\rho(0) = \rho(1) = 0$ and

$$\int_0^1 (\rho')^2 \, ds = 1 \quad ,$$

the work performed by the force μ must exceed the strain energy represented by the deflection $\rho(x)$, $x \in [0,1]$ (with the design $\underset{\sim}{s}(x)$). Specifically, for a given design $\underset{\sim}{s}$, $I(\underset{\sim}{s}(x))$ is a bounded piecewise-continuous (hence L_2 [0,1]) function. The constraints (4.3) generally imply that $I(\underset{\sim}{s}(x))$ is uniformly bounded for any choices of $\underset{\sim}{s}$ in the class of admissible designs.
We make the following observations (which are well known). Let $\underset{\sim}{s}$ be fixed.
The bilinear form $V(u,v) =$

$$\int_0^1 E\, I(\underset{\sim}{s}(x))\, u''(x,\underset{\sim}{s}(x))\, v''(x,\underset{\sim}{s}(x))dx$$

is symmetric, and coercive, i.e.,

$$V(u,v) = V(v,u),$$

$$V(u,v) \leq K_1 \,||\, u \,||\, ||\, v \,|| \,,$$

$$\text{for some } K_1 > 0 \,,$$

$$V(u,u) > K_2 \,||\, u \,|| \quad ,$$

$$\text{for some } K_2 > 0 \quad ,$$

$$\text{and for all } u,v \in H_0^1 \,[0,1]$$

The eigenvalue problem: find $\mu \in R$, $u \in H_0^1$ [0,1]
such that $V(u, v) = \mu\, W_0(u,v)$, for all $v \in H^1$,

(where $W_0(u,v) = \int_0^1 (u' \cdot v')dx)$,

has a discrete sequence of eigenvalue $\mu_1 \leq \mu_2 \leq \ldots$, and of corresponding eigenvectors.
Each eigenspace is of finite dimension. Coercivity of $V(u,v)$ is easy to establish if the space of admissible designs S excludes designs which contain singularities.
For example, if the only designs which are admissible were such that

$$I(\underset{\sim}{s}) \geq I_0 > 0 \text{ for all } x \in [0,1] \,,$$

for some number I_0, coercivity of the bilinear form $V(u,v)$ can be deduced directly form the definition.
Coercivity implies weak compactness in H_0^1, hence, one can deduce existence of weak limits for finite dimensional approximations. There are, of course, serious problems. Coercivity of V implies coercivity of the quadratic form

$$V - \mu W_0 = \frac{1}{2} \int_0^1 (EI(\underset{\sim}{s}(x))\, u''\, v'')dx$$

$$- \frac{\mu}{2} \int_0^1 (u' \cdot v')dx \text{ for small values of } \mu .$$

It does not imply either convergence or even existence of appropriate limit for this energy form if a sequence of designs $\underset{\sim}{s}_n$ converges to a design $\underset{\sim}{s}_0$ in the $L_2[0,1]$ norm.

The bifurcation phenomena provide clear counterexamples to possible theorems about weak compactness, uniqueness of solutions, or existence of optimal solutions for linear beam, plate, or shell operator eigenvalue problems restated as optimal design problems.

In this section we examine the behavior of the basic energy form V - W.

This analysis was carried out by Komkov in [24], but possible generalizations of the theory were not obvious for higher-dimensional problems.

However, embedding the problem in one higher dimension provided a way of testing for bifurcation phenomena in a novel manner. The energy difference V-W becomes "un-physical" for small changes of the domain in the vicinity of the critical design.

5. Energy criterion for the optimal design problem.

In the domain Ω we wish to solve the eigenvalue problem $A(h)w(\underset{\sim}{x}) = \Lambda_h B(h)w(\underset{\sim}{x})$

where in the one-dimensional case

$$A = \frac{d^2}{dx^2} (EI\ (h(x))\ \frac{d^2}{dx^2}\ )$$

and $B = \frac{d}{dx}$, or $B = \rho A(x)$ depending on the problem.

And in the two-dimensional case,

$$= D(x,y)\ \nabla^4(\cdot) + 2\frac{\partial D}{\partial x} \cdot \frac{\partial}{\partial x}\ \nabla^2(\cdot))$$

$$+ 2\frac{\partial D}{\partial y}\ \frac{\partial}{\partial y}\ \nabla^2(\cdot) + \nabla^2(D\nabla^2(\cdot))$$

$$-(1 - \nu)\ [\frac{\partial^2 D}{\partial y^2}\ \frac{\partial^2 \cdot}{\partial x^2} - 2\frac{\partial^2 D}{\partial x \partial y}\ \frac{\partial^2 \cdot}{\partial x \partial y} + \frac{\partial^2 D}{\partial x^2}\ \frac{\partial^2 \cdot}{\partial y^2}\ ], \text{ where}$$

$$D = D(x,y,h(x,y)) \quad , \text{ and}$$

$$B = \nabla^2 \quad , \text{ or } B = m(h(x,y)) .$$

More complicated operators occur in shell theory which is neglected in this presentation.

To find numerically "bimodality" phenomenon, one could follow the fairly obvious iteration scheme of finding a minimum for $J_1 = \int_0^{\ell} a(s(x))(y_1'')^2 dx$ in $H_0^2[0,\ell]$ with constraints $\int_0^\ell (y_1')^2 dx = 1$, and $\int_0^\ell \alpha P(s(x))\, dx = 1$, and for $J_2 = \int_0^\ell \alpha(s(x))\, (u_2'')^2 dx$ in $H_0^2[0,\ell]$ with constraints $\int_0^\ell \alpha(s(x))(y_1 y_2)\, dx = 0$, $\int_0^\ell (y_2')\, dx = 1$ where $\alpha\,(s(x))$ can be identified with $EI(s(x))$ in the most frequently encountered case.

Additional constraints could be assigned without changing the nature of the problem.

A numerical check is maintained on the magnitude of

$$\nabla\lambda = \int_0^\ell \alpha(s(x))\, [(y_1'')^2 - (y_2'')^2]\, dx$$

and the process is terminated when $|\nabla\lambda|$ is sufficiently small at some step during the iteration. A variation of the gradient projection method is employed to improve $J_1(s)$. The process is also terminated if at any step $J_1 > J_2$.

While this technique has the advantage of theoretical simplicity it is numerically very cumbersome, requiring many iterative steps and carries many potential sources of errors. In [24] the first author proposed an embedding technique for identification of bifurcation and bimodal points in the process of design optimization.

In one-dimensional case, this approach is easily comprehended if we regard length as additional parameter. First, let us fix the design (and the length) and establish an algorithm for determination of the deflection (i.e. state function) for the column. Let V-W denote the form

$$V\text{-}W = \int_0^\ell \phi(w(x), \underset{\sim}{h}(x))(w'')^2\, dx - \lambda \int_0^\ell (w')^2 dx.$$

Let $h(x)$ be given. If for a given value of λ there exists a trial function $\psi(x) \neq 0$, $\psi\; C^1[0,\ell]$ such that $\phi(\psi(x), \underset{\sim}{h}(x)) < 0$ then $\lambda_{\psi,h} = E \int_0^\ell I(h)(\psi'')^2 dx / \int_0^\ell (\psi')^2 dx$ is an upper bound on the first eigenvalue.

(Note: By imposing the condition $\int_0^\ell (\psi')^2 dx = 1$ we can identify V with the strain energy $V(\psi(h(x)), h(x))$.)

Let us restrict our problem to a finite dimensional subspace of $H_0^2\ \varepsilon[0,\ell]$ with orthonormal basis $\{\Phi_i\}$, $i = 1,2,\ldots,r$. Let us choose a trial function $\psi_0(x)$ such that $\psi_0'(x) = \Sigma c_i \Phi_i$, and choose an original design $h_0(x)$.

The condition $\int_0^\ell (\psi_0')^2 = 1$ becomes $\sum_{i=1}^r c_i^2 = 1$. We compute the energy form

$$V - W = \tfrac{1}{2} \int_0^\ell EI(h_0(x)) \cdot \Sigma(c_i \Phi_i')^2\, dx - \tfrac{1}{2} P(\Sigma\, c_i^2).$$

The "smooth" minimum of V-W corresponds to r-separate equations

$c_k\{E \int_0^\ell I(h_0(x))\, \Sigma\, c_i \Phi_i'(x)\, dx = 0$, i.e. either $c_k = 0$ or

$$5.1 \quad c_k = \frac{-\sum_{i\neq k} c_i \int_0^{\ell} I(h_0(k)\ \Phi_i'(x)\ \ dx}{\int_0^{\ell} (h_0(x)\ \Phi_k(x))\ dx}, \quad k = 1,2,\ldots,r.$$ subject to the normalizing

condition $\sum_{i=1}^{r} c_i^2 = 1$. This can be rewritten as a linear system

$$(5.2) \quad \frac{d}{dc_k}(V - W) = 0 \quad k = 1,2,\ldots,r \quad ,$$

$$\text{or } \sum \alpha_{ij} c_i = 0 \quad j = 1,2,\ldots r$$

with the same normalizing condition: $\sum_{i=1}^{r} c_i^2 = 1$. Here <,> denotes the usual

$L_2[o,\ell]$ product.

The equation (5.2) can be regarded as a constraint applied to the problem of shape optimization.

Hence, with this additional constraint condition, we can study the behavior of derivatives of V-W with respect to the design vector h(x) at the point $h = h_0(x)$. This is a routine procedure if V-W is differentiable. The main point of this section of our paper is to establish some indications warning the programmer about the lack of differentiability, or more specifically of the "bimodality" phenomenon. The obvious idea of trying local variations for $h_0(x)$ and testing if two vectors $\underset{\sim}{c}^i = \{c_1^i, c_2^i, \ldots, c_r^i\}$, i = 1,2 satisfy (4.5) and give "close" values of V-W leads to incredibly messy computational algorithms even in this simple one-dimensional case. In more realistic cases when the design vector may have hundreds of components it is best to abandon such obvious procedures before wasting computer time and a lot of programmer man-hours. In our case, a physically motivated warning seems to be readily apparent. We are considering buckling phenomena. The dependence of eigenvalue and eigenvectors on shape of the column, i.e. on h(x) is not easily established. However, a single scalar parameter ℓ (length) is a readily available indicator of bimodality. One needs to observe that ℓ is treated as a constant in our problem, but if one regards ℓ as a parameter and examines the functional dependence of V-W on ℓ, tell-tale signs are apparent that $h_0(x)$ (or the design of the structure in general is close to the critical bimodality point.) In [24] Komkov observed that in some neighborhood of the critical design the derivative of V-W with respect to ℓ changes its sign. To clarify this statement let us state a lemma.

Lemma: We consider the Euler-Bernoulli model of a column given by equations (1.1). Let U be the set of admissible designs. We assume that U is a bounded convex subset of a Banach space B. Let $h_0(x)$, $x \in [0,\ell]$, be a critical design, such that the corresponding state function W(x) minimizing the energy functional V-W is not unique.

Let $w_1(x)$, w_2 (x) be two state functions minimizing V-W over the Sobolev space H of weak solutions of (1.1). Then given $\varepsilon>0$ there exists a $\delta>0$ such that for any fixed $h(x)$ in ε-neighborhood of $h_0(x)$ (in V B) one can choose a suitable variation of ℓ of magnitude greater than δ for which a given $w_i(x)$ $(i = 1,2)$ minimizes V-W.

The proof uses the fact that the location of the critical point is a monotone function of ℓ for a fixed $h(x)$. The technicalities of the proof follow directly from Fourier series decomposition as given in [24].

The above discussion serves only the purpose of justifying the numerical algorithm that is given in the next section.

Energy criterion for the plate problem modelled by equation (5.2) is slightly more complicated.

However, the general idea of this technique is applicable. The single parameter h can be replaced by a number of dimensional parameters in the case of plate theory. In shell theory, one has a number of so-called "constants" that can be varied in this embedding technique.

Specific details of such embedding algorithm applied to higher dimensional problems or to more realistic nonlinear models have not been worked out in detail at this time.

We offer next an algorithm for the one dimensional case i.e., of column buckling for materials obeying Hooke's law. There is no basic conceptual difficulty preventing us from developing similar algorithms for higher dimensional problems or for more complex materials (i.e. materials satisfying different constitutive equations) and possibly for anisotropic materials.

A Numerical Algorithm. (Column Optimization)

1. For i = 0, choose S_0 = constant.
2. Subdivide $[o,\ell]$ into n equal parts.
3. Choose m.
4. Choose orthonormal functions $\Phi_1, \Phi_2, \ldots, \Phi_m$ obeying all specified boundary conditions.
5. Compute $a_1, a_2, \ldots, a_m$.
6. Compute numerically:

   $$V(S_0, \ell) = \frac{\pi^3}{4} \frac{S}{\ell^3} \sum_{n=1}^{m} n^4 a_n^2 - \sum_{n=1}^{m} a_n k_n(\ell) \rightarrow \min$$

7. $W(S_0,\ell) = \int_0^\ell q(x)\, \psi(x)dx$, where $\psi(x) = \sum_{n=1}^{m} a_n \Phi_n$.
8. Compute $\lambda_\psi = \dfrac{\frac{1}{2}\int_0^\ell S_0(x)(\psi'')^2 dx - W(S_c)}{\frac{1}{2}\int_0^\ell (\psi')^2\, dx}$
9. Compute the gradent $\dfrac{\partial\lambda\psi}{\partial S_i}$

10. Choose $S_i = S_o + \text{grad}\,(\lambda_\psi)\delta S_o$ where δS_o is chosen in the direction of -gradient S_i, and $||\delta S_i||<\delta_o$ is constraint on the size of change. Note: So far we have followed an established numerical procedure which can be found in papers of Haug, Roussellet, Choi, Arrora and others with only a slight variant consisting in selection of Fourier series coefficients, rather than direct optimization. Next steps are not found in any standard procedures and constitute the essential new check on bimodality.
11. Check the sensitivity of λ_ψ. If grad λ_ψ is "small", go to 14.
12. Check if V-W changes sign for small increment in ℓ.
13. If #12 is true that is the design is close to bimodal point, stop and print the results. If not, go to step #5 and repeat the loop.
14. Print S_i.
15. End.

We comment that step #12 of this algorithm is crucial in determining the uniqueness of the gradient procedure by eliminating the bimodality phenomenon.

Appendix I. Quasidifferentiability

Let $a\varepsilon B$, where B is a Banach space, and let $\Phi(a)$ be a (real-valued) functional Φ: B → R. Let us define directional derivative of Φ in the direction of $\eta\varepsilon B$ computed at $a = a_o$ in the usual manner.

$$\Phi_{a,\eta}|_{a=a_o} = \lim_{\varepsilon\to 0} \frac{\Phi(a_o + \varepsilon\eta) - \Phi(a_o)}{\varepsilon}$$

Let us suppose that the directional derivative exists for all choices of B.

We define a subdifferential $\partial\Phi$ of Φ at $a = a_o$ to be the set of all elements of B* given by the containment relation $m\varepsilon\partial\Phi|_{a=a_o}$ if $\Phi(a) - \Phi(a_o) \geq \langle m, (a-a_o)\rangle$ for all $a\varepsilon B$ in some neighborhood of a_o.

The following lemma has been proved in [14] and is discussed in the papers of Choi and Haug [15], and Haug and Rousselet [19]. Also see [32].

The subdifferential $\partial\Phi$ is convex, weak* closed and bounded in B*, under some fairly weak assumptions. Moreover, the directional derivative of Φ, i.e. $\Phi_{a\eta}|_{a=a_o}$ is equal to the supremum $\sup_{m\varepsilon\partial\Phi_{a=a_o}} (m(\eta))$.

Conversely, if one can find a weak* closed, bounded, convex neighborhood of a_o in B such that for any $\eta\varepsilon B$, it is true that $\Phi_{a,\eta}|_{a=a_o} = \sup_{m\varepsilon\partial\Phi_{a=a_o}} (m(\eta))$ then the functional Φ is called quasidifferentiable at a_o. For a detailed analysis of quasidifferentiability of eigenvalues in structural analysis, see [19]. The problem of interest to us is the failure of the iterative gradient projection method when it is applied to the lower eigenvalue, and the selection of a suitable subdifferential when the Fréchet derivative does not exist.

Appendix 2: Frechet and Gateaux Differentiation and Formal Rules of Fréchet Calculus

Let B_1 and B_2 be any normed spaces. A mapping $f : B_1 \to B_2$ is said to be Gateaux differentiable at $\bar{x} \in B_1$ if for any $h \in B_1$ and for any constant t such that $\bar{x} + th$ is in the domain of f, there exists a linear map f'(x) defined in some neighborhood of $\bar{x}$, $f'(x) : B_1 \to B_2$ such that

$$f(\bar{x} + th) - f(\bar{x}) = tf'(\bar{x})h + r(\bar{x},h,t)$$

where $\lim_{t \to 0} \frac{\|r\|}{t} = 0$ for all $h \in B_1$. In the Hilbert space setting (call it H), continuity of f' at $\bar{x}$ implies validity of the Riesz representation theorem for the specific case when $B_2 = \mathbb{R}$. Hence, in the case $f : H \to \mathbb{R}$, f'(x)h is an inner product, $f'(x)h = \langle z, h \rangle$. The operator z is called the Fréchet derivative of f. Higher order derivatives are defined analogously. If f and y are vectors in a Hilbert space then $\frac{\partial f}{\partial y}$ is a tensor product, and $\langle \frac{\partial f}{\partial y} x, x \rangle$ is a scalar. Thus, $\frac{\partial f}{\partial y}$ can be regarded as an operator from H to H. More specifically, if y(x) and x are n-dimensional vectors, then $\frac{\partial y}{\partial x}$ is an nxn matrix, which is in fact the Jacobian matrix.

The following rules of calculus are easily checked to be correct:

$$\frac{\partial}{\partial x}(\Phi_1 + \Phi_2) = \frac{\partial}{\partial x}\Phi_1 + \frac{\partial}{\partial x}\Phi_2$$

$$\frac{\partial}{\partial x}(c\Phi) = c\frac{\partial}{\partial x}\Phi \quad , \text{ where c is a constant,}$$

$$\frac{\partial}{\partial x}(uv) = u\frac{\partial v}{\partial x} + \frac{\partial u}{\partial x}v$$

$$\frac{\partial}{\partial x}\left(\frac{u}{v}\right) = v^{-2}\left(v\frac{\partial u}{\partial x} - u\frac{\partial v}{\partial x}\right), \text{ where v is a functional and } v^{-2} \text{ is defined}$$

$$\frac{\partial}{\partial x}\Phi(\psi(x)) = \frac{\partial \Phi}{\partial \psi}\cdot\frac{\partial \psi}{\partial x} \quad ,$$

where Φ is a functional $H_2 \to \mathbb{R}$, and ψ is a map : $H_1 \to H_2$, which itself is an element of a suitable Hilbert space. In particular, if A is a map $A : H \to H$, then

$$\frac{\partial}{\partial x}\langle Ax, x \rangle = Ax + A^*x \quad .$$

For a detailed exposition on the theory of Fréchet differentiation, see [27,28]. For applications to continuum mechanics, see [29].

For the general theory, see the monograph of Vaînberg [30] or the original work of Fréchet [31].

REFERENCES

1. Leonhardi Euleri Opera Omnia, vol.X, ser. secundae, Society for Natural Sciences of Switzerland, 1960, in particular the section of C. Truesdell, historical notes pp. 1638-1788, on the rational mechanics of flexible or elastic bodies.

2. J. L. Lagrange, Sur la figure des colonnes, Miscellanea Taurinensia, vol. V, 1970, (See pp. 123-125).

3. J. B. Keller, The shape of the strongest column, Archives of Rational Mechanics and Analysis, vol.5 (1960), pp. 275-285.

4. I. Tadjbakhsh and J.B. Keller, Strongest columns and isoperimetric inequalities for eigenvalues, J. of Applied Mechanics, vol. 9, (1962), pp. 159-164.

5. N. Olhoff, Optimal design against structural, vibration, and instability, Ph.D. Thesis, Technical University of Denmark, Dept. of Solid Mechanics, Lyngby, Denmark, November, 1978.

6. E.J. Haug, U.S. Army Material Command Pamphlet, AMC 706-902 (1972-73), Engineering Design Handbook.

7. F. Niordson and P. Pedersen, A review of optimal structural design, Applied Mechanics, Springer Verlag, Berlin, (1973), pp. 264-278.

8. N. Olhoff and S. H. Rasmussen, On single and bimodal optimum buckling loads of clamped columns, Int J. Solids & Structures, vol. B, (1979), pp. 605-614.

9. E. F. Masur and Z. Mróz, Nonstationary optimality conditions in structural design, Int J. Solids & Structures, vol. 15, (1979), pp. 503-512.

10. E. J. Haug and J. S. Arrora, Applied Optimal Design, J. Wiley - Interscience, New York, 1977.

11. E. J. Haug, K.C. Pan, and T.D. Streeter, A computational method for Optimal Structural Design, Part II, Continuous problems, J. Numerical Methods in Engineering, vol. 9, 1975, pp. 649-667.

12. J.E. Taylor and C.Y. Liu, Optimal Design of columns, AIAA Journal, vol. 6 (1968), pp. 1496-1502.

13. J.E. Taylor, The strongest column, An energy approach, J. Applied Mechanics, vol. 34 (1967), pp. 486-489.

14. B.N.Pshenichniĭ, Necessary conditions for an extremum, Marcel Dekker, New York, 1971, (Translated from Russian).

15. K.K. Choi and E.J. Haug, Repeated eigenvalues in mechanical optimization problems, Optimization of Distributed Parameter Structures, vol. I, Sijthoff & Noordhoff Publishers, the Netherlands, 1981, pp. 219 - 277.

16. E. Clarke, Generalized gradients and applications, Trans. Amer. Math. Society, 205, #2 (1975), pp. 247-262.

17. E.F. Masur, Optimality in the presence of discreteness and discontinuity, Proceedings IUTAM Symposium on Optimization in Structural Design, A. Sawczuk and Z. Mróz editors, Springer Verlag, Berlin, 1975, pp. 441-453.

18. G. H. Knightly and D. Sather, Buckled states of a spherical shell under uniform external pressure, Arch. Rat. Mechanics Anal. 72 (1890) pp. 315-380.

19. E.J. Haug and B. Rousselet, Design Sensitivity Analysis in Structural Mechanics, Part II, eigenvalue variations, J. Structural Mechanics, vol. 8, #2, 1980.

20. V. Komkov, An optimal design problem--a non-existence theorem, Arch.Mech., 33, (1981) pp. 147 - 151.

21. N. Olhoff, Optimization of vibrating beams with respect to higher order natural frequencies, J. Structural Mechanics vol. 4, (1976), pp. 87-122.

22. H. Rabitz, Sensitivity methods for mathematical modelling, this issue.

23. J. Tilden, V. Constanza, G. McRae and J. Seinfeld, Modelling in Chemical Reaction Systems, editors, K. Ebert, P. Deuflhard and J. Jaeger, Springer Verlag, 1981.

24. V. Komkov, An embedding technique in problems of elastic stability, ZAMM, 60, (1980), pp. 503-507.

25. E.J. Haug and J.S. Arora, Distributed Parameter Structural Optimization, vol. I, Sijthoff & Noordhoff Publishers, the Netherlands, 1981, pp. 219-277.

26. N.A. Shor, Minimization techniques for nondifferentiable functions, and their applications, Naukova Dumka, Kiev, 1979.

27. M.M. Vaǐnberg, Variational Methods for Investigation of Nonlinear Operators, (English Translation) Holden Day, San Francisco, 1963.

28. T. Kato, Perturbution Theory for Linear Operators, Springer Verlag, Berlin and New York, Die Grundlehren der mathematischen Wissenschaften series, vol. 132, 1966.

29. V. Komkov, On formulation of variational problems in the classical continuum mechanics of solids, International Journal of Engineering Science, vol.6, 1968, pp. 695-720.

30. M. M. Vainberg, On differentials and gradients of mappings, Uspekhi Matem. Nauk, 7, #49, (1952), pp. 139-143.

31. M. Fréchet, La notion de differentielle dans l'analyse generale. Ann. Soc. de l'Ecole Norm. Super., 42, (1925), pp. 293-323.

32. K.K. Choi, E.J. Haug, J.W. Hou and V.N. Sohoni, Pshenichnyi's Linearalization Method for Mechanical System Optimization, Trans. ASME, J. Mech. Design to appear in 1984.

SENSITIVITY ANALYSIS FOR NON-SELFADJOINT PROBLEMS

Pauli Pedersen

Department of Solid Mechanics
The Technical University of Denmark, Lyngby,
DK 2800, Denmark

ABSTRACT

The problems of stability for non-conservative systems are connected with the questions concerning stability of vibrations. These problems are described by non-selfadjoint operators or, alternatively, by non-symmetric matrices. The condition of stability depends on the parameters of the problem, i.e. on the design parameters, the boundary conditions, the load distribution, etc. Therefore it is important to obtain quantitative information about this dependence.

For these non-conservative problems it is shown, in general, how the different sensitivity analyses can be performed without introducing any new concepts of eigenvalue analysis. In the primary analysis as well as in the sensitivity analysis, the integrated treatment of the adjoint problem is of major importance because it admits a stationarity principle for these nonselfadjoint problems. Stated in another way, it means that if the variations are related to the "mutual energies", then the variations of eigenvector and adjoint eigenvector are not implicitly involved. One of the main questions asked in this paper relates to the change in the flutter load as a function of the change in stiffness, mass, boundary conditions or in the load distribution.

The extended "Beck column" is treated in a non-discretized analysis. Then we concentrate on the viscoelastic vibrating columns, which have to be treated by a discretized model. The results of our analysis, which clearly show the mutual effects of external and internal damping, are presented. Finally the sensitivities are computed for these rather complicated structural models.

Note

The theoretical results offered in this paper form a continuation of the previous work of the author and A.P. Seyranian, particularly of the work in reference [1].

1. INTRODUCTION

The notion of sensitivity analysis is a very general one. In this paper we restrict it to response sensitivity for problems of solid mechanics. To be more specific: we determine how do the structural displacements, stresses, eigenfrequencies, and stability of the load, change with a change in the parameters of the problem. As we shall see, a sensitivity analysis demands comparatively few calculations because all the necessary data are available from the primary analysis. However, for the non-selfadjoint (non-conservative) problems to be focused on these data, one must include the solution to the adjoint problem.

The usual (familiar) analysis for the displacements, etc., could be based either on the equations of equilibrium, set up directly, or on the stationarity of an energy functional. For the non-selfadjoint problems such functionals are now also available in the literature and we term them here the mutual energy functionals - mutual in the sense that displacements of the physical as well as of the adjoint problem are involved. The sensitivity analysis may also be based directly on the equations of equilibrium, but the simplicity is striking when we base it on the mutual energy, because then the variations of displacements do not have to be considered.

The response analysis and sensitivity analysis can be formulated in either a continuous or in a discrete form. As almost all problems are solved numerically with only a finite degree of freedom, the sensitivity analysis is presented in a matrix formulation, which has the advantage that boundary conditions are an integrated part. Furthermore, this formulation relates directly to the practical methods of analysis, such as the finite element method and the Galerkin method.

The most important results of the present paper are derived in the paper of PEDERSEN & SEYRANIAN [1], but the presentation here is somewhat different, and an extended discussion with relation to finite element and Galerkin discretizations is included. Also a new discussion relating to the case of pure external damping is presented, and a stabilization theorem is proved.

A suitable list of references is given in [1] and also can be found in SEYRANIAN [2], where the author choses a continuous operator formulation.

This paper is an enlarged version of the talk given by the author at the American Mathematical Society meeting in New York City on April 15-th, 1983.

See the Notices of the American Mathematical Society, March, 1983.

2. VARIATIONAL ANALYSIS

In principle we are interested in studying the dynamic behaviour of statically loaded systems (structures). However, the time dimension τ is separated by the exponential function

$$e^{\lambda\tau} = e^{(\alpha+i\omega)\tau} = e^{\alpha\tau}e^{i\omega\tau} = e^{\alpha\tau}(\cos\omega\tau + i\sin\omega\tau) \quad , \tag{2.1}$$

following which, the spatial problem (vibration mode) is described by the homogeneous matrix equation

$$[L]\{\Phi\} = \{0\} \quad \text{or} \quad \{\Phi\}^T[L]^T = \{0\}^T \ , \tag{2.2}$$

where the system matrix $[L]$ depends on the complex eigenvalue $\lambda = \alpha + i\omega$, with α as a stability measure and ω as frequency according to (2.1). Furthermore, $[L]$ depends on the load level described by the real parameter p, and on design, load distribution, damping, etc., all of which we symbolize by the real quantity h, which - in the individual cases - may be a scalar parameter or a spatial parameter function $h(x)$.

In addition to the physical problem (2.2), we analyse the adjoint problem

$$[L]^T\{\Psi\} = \{0\} \quad \text{or} \quad \{\Psi\}^T[L] = \{0\}^T \ . \tag{2.3}$$

Note that the eigenvectors $\{\Phi\}$ of (2.2) and $\{\Psi\}$ of (2.3) are generally complex eigenvectors.

A functional Π, which may be interpreted as a mutual potential, is defined by

$$\Pi := \{\Psi\}^T[L]\{\Phi\} = \{\Phi\}^T[L]^T\{\Psi\} = 0 \ , \tag{2.4}$$

where the zero follows from (2.2) or (2.3). Taking general variations of Π, we get, again using (2.2) and (2.3),

$$\delta\Pi = \{\delta\Psi\}^T[L]\{\Phi\} + \{\Psi\}^T[\delta L]\{\Phi\} + \{\Psi\}^T[L]\{\delta\Phi\} = \{\Psi\}^T[\delta L]\{\Phi\} = 0 \ . \tag{2.5}$$

From (2.5) with unchanged system matrix $[\delta L] = [0]$, we read that Π is stationary with respect to eigenvector variations or, alternatively, that equilibrium (2.2) and (2.3) follow from stationarity of Π with respect to arbitrary variations $\{\delta\Phi\}$, $\{\delta\Psi\}$.

In relation to sensitivity analysis, where $[\delta L] \neq [0]$, the main conclusion to be drawn from (2.5) is that eigenvector variations $\{\delta\Phi\}$ and $\{\delta\Psi\}$ are not involved. Thus, writing (2.5) more specifically in the variations of the involved parameters λ, p and h, we simply have

$$\{\Psi\}^T[L]_{,\lambda}\{\Phi\}\delta\lambda + \{\Psi\}^T[L]_{,p}\{\Phi\}\delta p + \{\Psi\}^T[L]_{,h}\{\Phi\}\delta h = A\delta\lambda + B\delta p + C\delta h = 0 \ , \tag{2.6}$$

with $[L]_{,\lambda}$ as symbol for the matrix obtained by partial differentiation with respect to λ of all matrix elements. For later convenience we have defined the complex quantities A, B and C according to (2.6) by

$$A := \{\Psi\}^T[L]_{,\lambda}\{\Phi\} \ , \ B := \{\Psi\}^T[L]_{,p}\{\Phi\} \ , \ C := \{\Psi\}^T[L]_{,h}\{\Phi\} \ . \tag{2.7}$$

Now, different but all direct interpretations of (2.6) give the results needed.

Sensitivities with respect to change in load level. Even with an unchanged "design", i.e. $\delta h = 0$ the sensitivity analysis is important and makes possible a more rigorous definition of terms like critical load and instability load.

The degenerated case of $A = 0$ will be discussed later, and with $A \neq 0$ and $\delta h = 0$ we read from (2.6):

$$\frac{\partial\lambda}{\partial p} = -\frac{B}{A} \quad \text{or} \quad \frac{\partial\alpha}{\partial p} = -\,\mathcal{Re}\left(\frac{B}{A}\right) \; , \quad \frac{\partial\omega}{\partial p} = -\,\mathcal{Im}\left(\frac{B}{A}\right) . \tag{2.8}$$

A critical load level is normally defined by the condition $\alpha = 0$ (periodic time function (2.1)). To clarify the term flutter load level p_F, we further require $\partial\alpha/\partial p > 0$, i.e. a sensitivity is involved. Using (2.8), we therefore have

$$p = p_F \Rightarrow \lambda = \lambda_F = i\omega_F \; , \quad \mathcal{Re}(B/A) < 0 . \tag{2.9}$$

The divergence case is included by means of $\omega_F = 0$.

Sensitivities with respect to change in design. Other important sensitivities are related to the change in behaviour $\delta\lambda = \delta\alpha + i\delta\omega$ due to changes in design, load distribution, damping, etc., but at unchanged load level. For this case with $A \neq 0$ and $\delta p = 0$ we read from (2.6):

$$\frac{\partial\lambda}{\partial h} = -\frac{C}{A} \quad \text{or} \quad \frac{\partial\alpha}{\partial h} = -\,\mathcal{Re}\left(\frac{C}{A}\right) \; , \quad \frac{\partial\omega}{\partial h} = -\,\mathcal{Im}\left(\frac{C}{A}\right) . \tag{2.10}$$

We are often interested in knowing whether a certain parameter h acts in a stabilizing or a destabilizing manner, which, from (2.10), we write as

$$\begin{aligned} \mathcal{Re}(C/A) > 0 &\Leftrightarrow h \text{ is stabilizing,} \\ \mathcal{Re}(C/A) < 0 &\Leftrightarrow h \text{ is destabilizing.} \end{aligned} \tag{2.11}$$

An example where h is a damping parameter will be discussed later.

Sensitivities of the flutter load level. The main question of the present paper relates to change in instability level as a function of design, i.e. all variations of (2.6) are involved. Since we focus on a load level of initial instability, the variations are restricted by

$$\alpha = 0 \; , \quad \delta\alpha = 0 \; , \quad \text{i.e.} \quad \lambda = i\omega_F \; , \quad \delta\lambda = i\delta\omega_F \; , \tag{2.12}$$

with divergence as the specific case of $\omega_F = \delta\omega_F = 0$. Thus, (2.6) gives

$$A(i\delta\omega_F) + B\delta p_F + C\delta h = 0 \; , \tag{2.13}$$

with the solution for $A \neq 0$ and from (2.9) $\mathcal{Re}(B/A) < 0$,

$$\partial p_F/\partial h = -\,\mathcal{Re}(C/A)/\mathcal{Re}(B/A) . \tag{2.14}$$

Note that the sign of $\partial p_F/\partial h$ is equal to the sign of $\mathcal{Re}(C/A)$. This simply states the natural fact from (2.11) that if h is stabilizing, then p_F increases with h. From (2.8) and (2.10) we may also write

$$\partial p_F/\partial h = -\left((\partial\alpha/\partial h)/(\partial\alpha/\partial p)\right)_{p = p_F} . \tag{2.15}$$

Sensitivities of the flutter frequency. The change in flutter frequency is also obtained from (2.13) by

(2.16) $$\partial\omega_F/\partial h = - Im(C/B)/Re(A/B) ,$$

because δp_F is a real quantity. That $B \neq 0$ follows from the ineq. (2.9).

3. AN IMPORTANT CLASS OF PROBLEMS

We restrict this section to problems described by a system matrix $[L]$, where the dependence on p and λ are shown explicitly

(3.1) $$[L] = [S] + p[K] + \lambda[C] + \lambda^2[M] ,$$

with the right-hand side matrices all being real.

In order to get a convenient short notation we define "specific" mutual elastic U, kinetic T, dissipative D, and external W energies by

(3.2) $$U := \{\Psi\}^T[S]\{\Phi\} ,$$

(3.3) $$T := \{\Psi\}^T[M]\{\Phi\} ,$$

(3.4) $$D := \{\Psi\}^T[C]\{\Phi\} ,$$

(3.5) $$W := - \{\Psi\}^T[K]\{\Phi\} ,$$

and if equilibrium (2.2) is used we may express the specific mutual external energy by

(3.6) $$W = (U + \lambda D + \lambda^2 T)/p .$$

The important complex quantities A and B of section two for the system matrix (3.1) will then be

(3.7) $$A = \{\Psi\}^T[L]_{,\lambda}\{\Phi\} = D + 2\lambda T ,$$

(3.8) $$B = \{\Psi\}^T[L]_{,p}\{\Phi\} = \{\Psi\}^T[K]\{\Phi\} = - W = - (U + \lambda D + \lambda^2 T)/p ,$$

and for the important sensitivity to pure load change (2.8) we get

(3.9) $$\frac{\partial\lambda}{\partial p} = \frac{\partial\alpha}{\partial p} + i\,\frac{\partial\omega}{\partial p} = \frac{W}{(D + 2\lambda T)} = \frac{U + \lambda D + \lambda^2 T}{p(D + 2\lambda T)} .$$

With regard to design sensitivity, let h be a parameter without influence on $[K]$; then, complex C of section two is

(3.10) $$C = \{\Psi\}^T[L]_{,h}\{\Phi\} = \{\Psi\}^T[S]_{,h}\{\Phi\} + \lambda\{\Psi\}^T[C]_{,h}\{\Phi\} + \lambda^2\{\Psi\}^T[M]_{,h}\{\Phi\} .$$

Finite element modelling. The expression (3.10) seems rather complicated, but a closer look in relation to the practical methods of usual analysis will show the simplicity. Let the system matrices be set up by the finite element method (FEM); then, $[S]$, $[C]$ and $[M]$ are obtained by accumulation over all elements

(3.11) $$[S] = \sum_e [S_e] \quad , \quad [C] = \sum_e [C_e] \quad , \quad [M] = \sum_e [M_e] \; .$$

This has to be read symbolically, because $[S]$, $[C]$ and $[M]$ are higher order matrices, say of order 10^3 , while $[S_e]$, $[C_e]$ and $[M_e]$ are lower order matrices, say of order 10 . Often, a parameter h_e , say the thickness of element e , only influences that specific element, and then (3.10) is brought down to the lower order

(3.12) $$C = \{\Psi_e\}^T[S_e]_{,h_e}\{\Phi_e\} + \lambda\{\Psi_e\}^T[C_e]_{,h_e}\{\Phi_e\} + \lambda^2\{\Psi_e\}^T[M_e]_{,h_e}\{\Phi_e\} \; ,$$

where the elements of the lower order vectors $\{\Psi_e\}$, $\{\Phi_e\}$ are contained in the higher order vectors $\{\Psi\}$, $\{\Phi\}$.

Furthermore, the h_e dependence is often homogeneous (extension to polynomial dependence is straight forward), such that

(3.13) $$[S_e]_{,h_e} = k[S_e]/h_e \quad , \quad [C_e]_{,h_e} = \ell[C_e]/h_e \quad , \quad [M_e]_{,h_e} = m[M_e]/h_e \; ,$$

where k , ℓ , m are integers, and then (3.12) can be written

(3.14) $$C = \left(\{\Psi_e\}^T[S_e]\{\Phi_e\}k + \lambda\{\Psi_e\}^T[C_e]\{\Phi_e\}\ell + \lambda^2\{\Psi_e\}^T[M_e]\{\Phi_e\}m\right)/h_e$$

$$= (kU_e + \ell\lambda D_e + m\lambda^2 T_e)/h_e \; ,$$

defining the element mutual energies which from (3.11) and (3.2)-(3.4) by accumulation gives the total energies

(3.15) $$U = \sum_e U_e \quad , \quad D = \sum_e D_e \quad , \quad T = \sum_e T_e \; .$$

For the result (3.14), the design sensitivities (2.10) will be

(3.16) $$\frac{\partial\lambda}{\partial h_e} = \frac{\partial\alpha}{\partial h_e} + i\,\frac{\partial\omega}{\partial h_e} = \frac{-\,(kU_e + \ell\lambda D_e + m\lambda^2 T_e)}{h_e(D + 2\lambda T)} \; .$$

The flutter load sensitivity (2.14) is then

(3.17) $$\frac{\partial p_F}{\partial h_e} = \frac{p_F}{h_e}\,\frac{\mathcal{R}e((kU_e + \ell\lambda D_e + m\lambda^2 T_e)/(D + 2\lambda T))}{\mathcal{R}e((U + \lambda D + \lambda^2 T)/(D + 2\lambda T))}$$

$$= \frac{1}{h_e}\,\frac{\mathcal{R}e((kU_e + \ell\lambda D_e + m\lambda^2 T_e)/(D + 2\lambda T))}{\mathcal{R}e(W/(D + 2\lambda T))} \; ,$$

and the flutter frequency sensitivity (2.16) is

(3.18) $$\frac{\partial\omega_F}{\partial h_e} = \frac{-1}{h_e}\,\frac{\mathcal{I}m((kU_e + \ell\lambda D_e + m\lambda^2 T_e)/(U + \lambda D + \lambda^2 T))}{\mathcal{R}e((D + 2\lambda T)/(U + \lambda D + \lambda^2 T))}$$

$$= \frac{-1}{h_e}\,\frac{\mathcal{I}m((kU_e + \ell\lambda D_e + m\lambda^2 T_e)/W)}{\mathcal{R}e((D + 2\lambda T)/W)} \; .$$

Global expansion modelling. An alternative to the FEM is the Galerkin, Ritz, Weighted Residual, etc. discretizations. In these methods the system matrices are obtained by integration over the structural domain V , with given expansion functions $u_i = u_i(x)$ for the physical and $v_j = v_j(x)$ for the adjoint eigenfunctions. Let us rewrite (3.10) without matrix notation

$$C = \sum_{ij}\sum (s_{ij,h} + \lambda c_{ij,h} + \lambda^2 m_{ij,h}) \psi_i \phi_j \ , \tag{3.19}$$

where s_{ij} , c_{ij} , m_{ij} are matrix elements of [S] , [C] , [M] , respectively. These elements are determined by

$$\begin{aligned} s_{ij} &= \int_V s\, F(u_i , v_j) dx \ , \\ c_{ij} &= \int_V c\, G(u_i , v_j) dx \ , \\ m_{ij} &= \int_V m\, H(u_i , v_j) dx \ , \end{aligned} \tag{3.20}$$

with $F(u , v)$, $G(u , v)$ and $H(u , v)$ as given expressions in the functions u , v and their spatial derivatives. The factors $s = s(x)$, $c = c(x)$ and $m = m(x)$ depend on the design function $h = h(x)$. By the partial derivative with respect to $h(x)$ we mean change of h at position x in space, and assuming only local influence, we write

$$\begin{aligned} s_{ij,h(x)} &= s_{,h(x)} F(u_i , v_j) \ , \\ c_{ij,h(x)} &= c_{,h(x)} G(u_i , v_j) \ , \\ m_{ij,h(x)} &= m_{,h(x)} H(u_i , v_j) \ . \end{aligned} \tag{3.21}$$

Inserting this in (3.19) we have

$$C(x) = \sum_{ij}\sum \Big(s_{,h(x)} F(u_i , v_j) + \lambda c_{,h(x)} G(u_i , v_j) + \lambda^2 m_{,h(x)} H(u_i , v_j) \Big) \psi_i \phi_j \ , \tag{3.22}$$

and the sensitivities (2.10), (2.14) and (2.16) are then a function of space. We often term them gradient functions $g(x)$, for example in relation to (2.14),

$$g(x) = \frac{\partial p_F}{\partial h(x)} = - \mathcal{R}e(C(x)/A)/\mathcal{R}e(B/A) \ , \tag{3.23}$$

and the resulting flutter load variation is obtained by integration

$$\delta p_F = \int_V g(x)\, \delta h(x)\, dx \ . \tag{3.24}$$

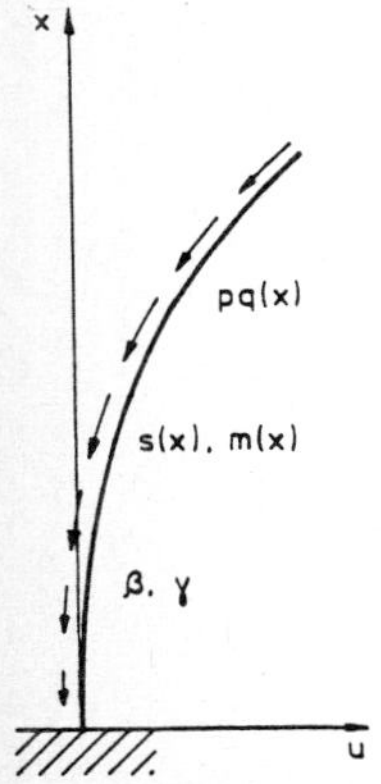

Fig. 4.1: Extended Beck/Leipholz/ Hauger columns.

4. EXAMPLES

Before concentrating on the flutter load sensitivity, let us study the results of an usual analysis for the Beck/Leipholz/Hauger columns shown in fig. 4.1. The characteristic curves $\omega = \omega(p)$ and $\alpha = \alpha(p)$ related to the first two modes are shown in fig. 4.2 for the case of a uniform Hauger column with $s(x) = m^2(x)$, $m(x) \equiv 1$, without internal damping $\gamma = 0$, but with external damping by the real non-negative parameter β .

The principal look of this result is independent of β and the load distribution, i.e. it is also valid for Beck, Leipholz columns. Thus, the interesting question is whether we can predict from the sensitivity analysis the constant solution $\alpha = -\beta/(2m)$ over the large load domain $0 \leq p \leq p_B$, where p_B is the load level at which we have a bimodal frequency. Note that only for $\beta = 0$ is p_B the flutter load, so it is normally a stable load level.

To obtain fig. 4.2 we have, as described in detail in [1], discretized the continuous problem and obtained the elements of [S] , [K] , [C] and [M] by

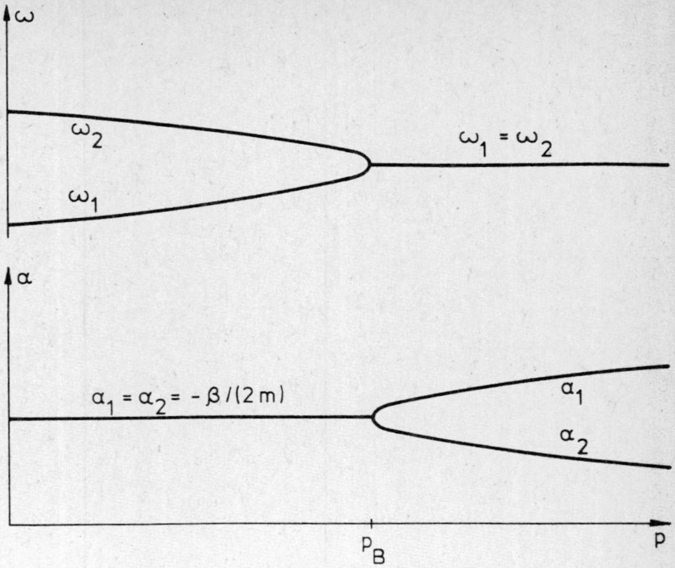

Fig. 4.2: Characteristic curves for the Hauger column $(q(x) = 1 - x)$.

$$s_{ij} = \int_0^1 s\, v_i'' u_j'' dx \quad , \quad k_{ij} = \int_0^1 \left(\int_x^1 q(\xi) d\xi \right) v_i u_j'' dx \quad , \tag{4.1}$$

$$c_{ij} = \beta \int_0^1 v_i u_j dx + \gamma s_{ij} \quad , \quad m_{ij} = \int_0^1 m\, v_i u_j dx \quad .$$

For the specific case of $m(x) \equiv m$ and $\gamma = 0$, this simplifies to

$$[C] = \frac{\beta}{m}[M] \Rightarrow D = \frac{\beta}{m}T \quad , \tag{4.2}$$

and then the system matrix [L] is

$$[L] = [S] + p[K] + (\beta\lambda/m + \lambda^2)[M] \quad , \tag{4.3}$$

and the sensitivity with respect to pure load level change (3.9)

$$\frac{\partial \lambda}{\partial p} = \frac{\partial \alpha}{\partial p} + i \frac{\partial \omega}{\partial p} = \frac{U/T + (\beta\lambda/m + \lambda^2)}{p(\beta/m + 2\lambda)} \quad . \tag{4.4}$$

Let us then <u>postulate</u> the solution $\alpha = -\beta/(2m)$, which makes $(\beta\lambda/m + \lambda^2)$ real, i.e.

$$\lambda = -\frac{\beta}{2m} + i\omega \Rightarrow \frac{\beta\lambda}{m} + \lambda^2 = -\left(\omega^2 + \frac{\beta^2}{4m^2}\right) \quad \text{and} \quad \frac{\beta}{m} + 2\lambda = i2\omega \quad . \tag{4.5}$$

Eigenvectors $\{\Phi\}$ and $\{\Psi\}$ will then be real, and U , T and U/T will thus also be real. In relation to (4.4) this gives

$$\frac{\partial \alpha}{\partial p} = 0 \quad , \quad \frac{\partial \omega}{\partial p} = \frac{1}{2\omega p}\left(\omega^2 + \frac{\beta^2}{4m^2} - \frac{U}{T}\right) \quad , \tag{4.6}$$

and thus proves unchanged $\alpha = -\beta/(2m)$. The increase in ω_1 and decrease in ω_2 as seen in fig. 4.2, according to (4.6) is due to the different energy ratios U/T for the two modes.

How is this simple behaviour (4.6) stopped at $p = p_B$? To understand this we return to the preassumption for (2.8), and note that for the present problem, $A \neq 0$ only for $T \neq 0$. We then prove $T = 0$ at p_B, i.e. when $\omega_1 = \omega_2$. For $p < p_B$, let ω_i be the frequency corresponding to $\{\Phi\}_i$; then, from (2.2), (4.3), (4.5) and premultiplication by $\{\Psi\}_j$, we have

(4.7) $$\{\Psi\}_j^T[S]\{\Phi\}_i + p\{\Psi\}_j^T[K]\{\Phi\}_i - \left(\omega_i^2 + \frac{\beta^2}{4m^2}\right)\{\Psi\}_j^T[M]\{\Phi\}_i = 0 .$$

Analogously, for ω_j, $\{\Psi\}_j$ and eq. (2.3), after postmultiplication by $\{\Phi\}_i$, we have

(4.8) $$\{\Psi\}_j^T[S]\{\Phi\}_i + p\{\Psi\}_j^T[K]\{\Phi\}_i - \left(\omega_j^2 + \frac{\beta^2}{4m^2}\right)\{\Psi\}_j^T[M]\{\Phi\}_i = 0 .$$

Eq. (4.7) minus eq. (4.8) gives

(4.9) $$(\omega_j^2 - \omega_i^2)\{\Psi\}_j^T[M]\{\Phi\}_i = 0 ,$$

which, for $\omega_j^2 \neq \omega_i^2$, shows what, for undamped systems, is termed the bi-orthogonality condition:

(4.10) $$\{\Psi\}_j^T[M]\{\Phi\}_i = 0 \quad \text{for} \quad \omega_j^2 \neq \omega_i^2 .$$

Then, by arguments of continuity, at the bimodal point, we get

(4.11) $$\{\Psi\}_B^T[M]\{\Phi\}_B = T_B = 0 ,$$

which, for undamped systems $(\beta = 0 \Rightarrow p_B = p_F ,\ \{\Psi\}_B = \{\Psi\}_F ,\ \{\Phi\}_B = \{\Phi\}_F)$, is normally termed the flutter condition. Now, the preassumption of (4.6) at $p = p_B$ is violated by (4.11), and the postulated solution (4.5) no longer holds, as seen in fig. 4.2.

For the same problem, let us see how we can prove β to be stabilizing. Therefore, let the "design" parameter h be equal to β, and we get, with (4.3),

(4.12) $$C = \{\Psi\}^T[L]_{,\beta}\{\Phi\} = (\lambda/m)T ,$$

and with $A = (\beta/m + 2\lambda)T$ also inserted in (2.11),

(4.13) $$Re\left(\frac{C}{A}\right) = Re\left(\frac{\lambda}{\beta + 2m\lambda}\right) = \frac{\beta\alpha + 2m(\alpha^2 + \omega^2)}{(\beta + 2m\alpha)^2 + 4m^2\omega^2} ,$$

which surely at $(\alpha = 0)$ and above $(\alpha > 0)$ the flutter load is positive, and thus proves that β is stabilizing. As stated in relation to (2.15), we then also have $\partial p_F/\partial\beta > 0$.

Finally, we shall show some numerical results for the complicated problems of non-uniform columns $m = m(x)$, $s(x) = m^2(x)$, and with internal damping $\gamma \neq 0$. Results from usual analyses are given in fig. 4.3, and flutter load gradient functions are shown in fig. 4.4.

The expression for this specific gradient function is evaluated from (3.23), (3.22), (3.8), and (3.7), with (4.1), and the design parameter h being equal to m $(s = m^2)$, we have

(4.14) $$g(x) = \frac{\partial p_F}{\partial m(x)} = \frac{Re\left(\left(\sum_{ij}\left((1 + \lambda\gamma)\,2mv_i''u_j'' + \lambda^2 v_i u_j\right)\psi_i\phi_j\right)/(D + 2\lambda T)\right)}{Re(W/(D + 2\lambda T))} .$$

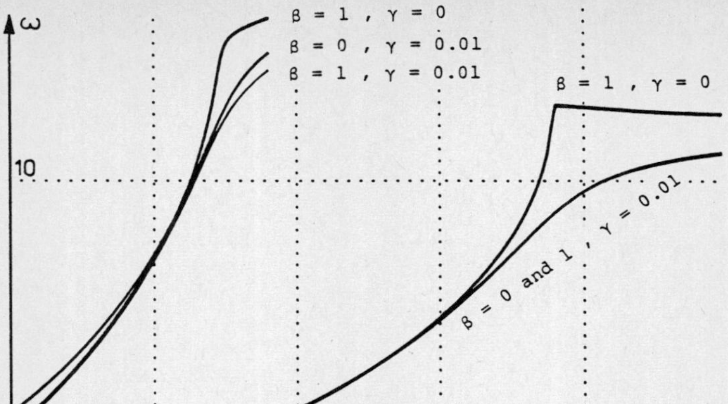

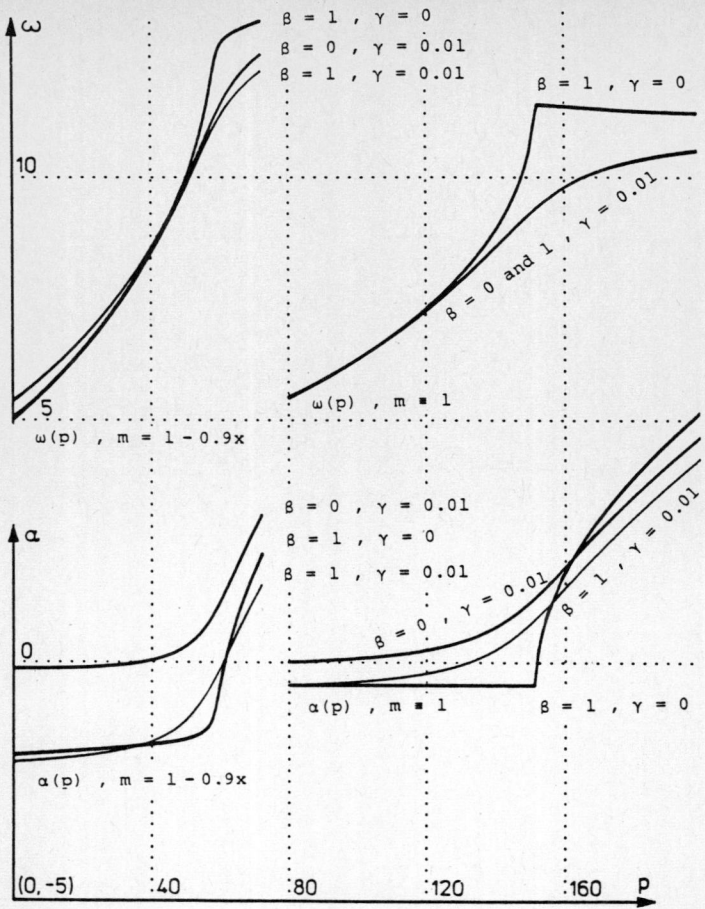

Fig. 4.3: Characteristic curves (ω_2 , α_2 omitted) for the Hauger column with external and internal damping, curves to the right of a uniform column, and curves to the left of a linearly tapered column.

Before ending the paper, the wide aspects of sensitivity analyses may be illustrated by determining the gradient function related to change of load distribution. Let the all-round parameter h now be q and thus only influencing the $[K]$ matrix. Analogously to (3.22) we get

$$C(x) = \sum_{ij}\sum k_{,q(x)}\, E(u_i , v_j)\psi_i\phi_j \ , \tag{4.15}$$

with $k = p_F\int_x^1 q(\xi)d\xi$ and $E(u_i , v_j) = v_i u_j''$ from (4.1). Then using integration by parts we get

$$C(x) = p_F \sum_{ij}\sum \int_0^x v_i(\xi)u_j''(\xi)d\xi\psi_i\phi_j \ , \tag{4.16}$$

and the resulting gradient is

$$g(x) = \frac{\partial p_F}{\partial q(x)} = \frac{p_F Re\left(\left(\sum_{ij}\sum \int_0^x v_i(\xi)u_j''(\xi)d\xi\psi_i\phi_j\right)/(D + 2\lambda T)\right)}{Re(W/(D + 2\lambda T))} \ . \tag{4.17}$$

The continuous analogue to this is given in [1].

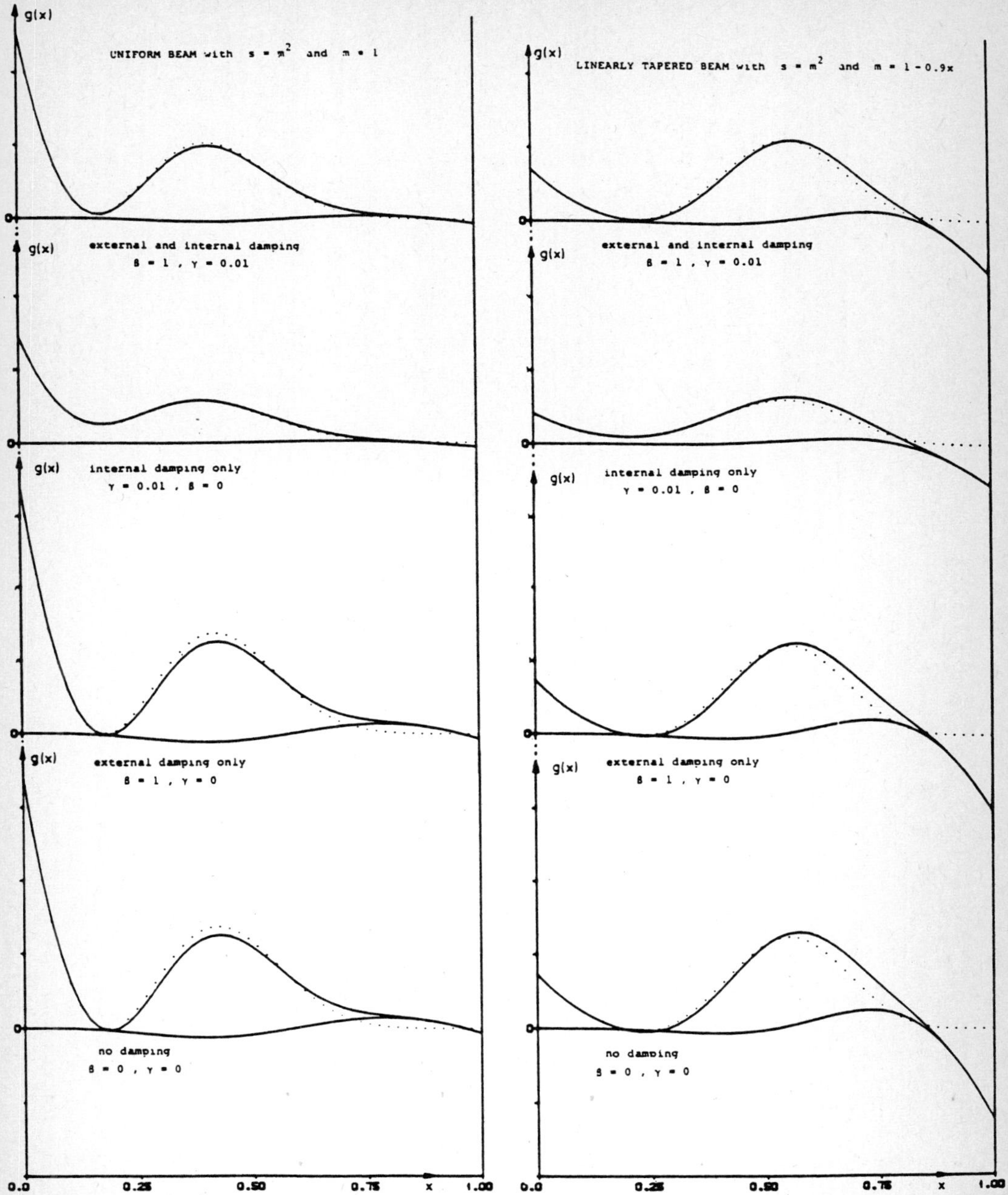

Fig. 4.4: Gradient functions (4.14) for uniform and linearly tapered Hauger columns with different external and internal damping. The influence of non-structural masses (term $\lambda^2 v_i u_j$) are shown separately.

REFERENCES

[1] Pedersen, P. and Seyranian, A.P.: Sensitivity Analysis for Problems of Dynamic Stability, Int. J. Solids Structures, Vol. 19, No. 4, pp. 315-335, 1983.

[2] Seyranian, A.P.: Sensitivity Analysis and Optimization of Aeroelastic Stability Characteristics, Int. J. Solids Structures, Vol. 18, No. 9, pp. 791-808, 1982.

Vol. 926: Geometric Techniques in Gauge Theories. Proceedings, 1981. Edited by R. Martini and E.M.de Jager. IX, 219 pages. 1982.

Vol. 927: Y. Z. Flicker, The Trace Formula and Base Change for GL (3). XII, 204 pages. 1982.

Vol. 928: Probability Measures on Groups. Proceedings 1981. Edited by H. Heyer. X, 477 pages. 1982.

Vol. 929: Ecole d'Eté de Probabilités de Saint-Flour X - 1980. Proceedings, 1980. Edited by P.L. Hennequin. X, 313 pages. 1982.

Vol. 930: P. Berthelot, L. Breen, et W. Messing, Théorie de Dieudonné Cristalline II. XI, 261 pages. 1982.

Vol. 931: D.M. Arnold, Finite Rank Torsion Free Abelian Groups and Rings. VII, 191 pages. 1982.

Vol. 932: Analytic Theory of Continued Fractions. Proceedings, 1981. Edited by W.B. Jones, W.J. Thron, and H. Waadeland. VI, 240 pages. 1982.

Vol. 933: Lie Algebras and Related Topics. Proceedings, 1981. Edited by D. Winter. VI, 236 pages. 1982.

Vol. 934: M. Sakai, Quadrature Domains. IV, 133 pages. 1982.

Vol. 935: R. Sot, Simple Morphisms in Algebraic Geometry. IV, 146 pages. 1982.

Vol. 936: S.M. Khaleelulla, Counterexamples in Topological Vector Spaces. XXI, 179 pages. 1982.

Vol. 937: E. Combet, Intégrales Exponentielles. VIII, 114 pages. 1982.

Vol. 938: Number Theory. Proceedings, 1981. Edited by K. Alladi. IX, 177 pages. 1982.

Vol. 939: Martingale Theory in Harmonic Analysis and Banach Spaces. Proceedings 1981. Edited by J.-A. Chao and W.A. Woyczyński. VIII, 225 pages. 1982.

Vol. 940: S. Shelah, Proper Forcing. XXIX, 496 pages. 1982.

Vol. 941: A. Legrand Homotopie des Espaces de Sections. VII, 132 pages. 1982.

Vol. 942: Theory and Applications of Singular Perturbations. Proceedings, 1981. Edited by W. Eckhaus and E.M. de Jager. V, 363 pages. 1982.

Vol. 943: V. Ancona, G. Tomassini, Modifications Analytiques. IV, 120 pages. 1982.

Vol. 944: Representations of Algebras. Workshop Proceedings, 1980. Edited by M. Auslander and E. Lluis. V, 258 pages. 1982.

Vol. 945: Measure Theory. Oberwolfach 1981, Proceedings. Edited by D. Kölzow and D. Maharam-Stone. XV, 431 pages. 1982.

Vol. 946: N. Spaltenstein, Classes Unipotentes et Sous-groupes de Borel. IX, 259 pages. 1982.

Vol. 947: Algebraic Threefolds. Proceedings, 1981. Edited by A. Conte. VII, 315 pages. 1982.

Vol. 948: Functional Analysis. Proceedings, 1981. Edited by D. Butković, H. Kraljević, and S. Kurepa. X, 239 pages. 1982.

Vol. 949: Harmonic Maps. Proceedings, 1980. Edited by R.J. Knill, M. Kalka and H.C.J. Sealey. V, 158 pages. 1982.

Vol. 950: Complex Analysis. Proceedings, 1980. Edited by J. Eells. IV, 428 pages. 1982.

Vol. 951: Advances in Non-Commutative Ring Theory. Proceedings, 1981. Edited by P.J. Fleury. V, 142 pages. 1982.

Vol. 952: Combinatorial Mathematics IX. Proceedings, 1981. Edited by E. Billington, S. Oates-Williams, and A.P. Street. XI, 443 pages. 1982.

Vol. 953: Iterative Solution of Nonlinear Systems of Equations. Proceedings, 1982. Edited by R. Ansorge, Th. Meis, and W. Törnig. VII, 202 pages. 1982.

Vol. 954: S.G. Pandit, S.G. Deo, Differential Systems Involving Impulses. VII, 102 pages. 1982.

Vol. 955: G. Gierz, Bundles of Topological Vector Spaces and Their Duality. IV, 296 pages. 1982.

Vol. 956: Group Actions and Vector Fields. Proceedings, 1981. Edited by J.B. Carrell. V, 144 pages. 1982.

Vol. 957: Differential Equations. Proceedings, 1981. Edited by D.G. de Figueiredo. VIII, 301 pages. 1982.

Vol. 958: F.R. Beyl, J. Tappe, Group Extensions, Representations, and the Schur Multiplicator. IV, 278 pages. 1982.

Vol. 959: Géométrie Algébrique Réelle et Formes Quadratiques, Proceedings, 1981. Edité par J.-L. Colliot-Thélène, M. Coste, L. Mahé, et M.-F. Roy. X, 458 pages. 1982.

Vol. 960: Multigrid Methods. Proceedings, 1981. Edited by W. Hackbusch and U. Trottenberg. VII, 652 pages. 1982.

Vol. 961: Algebraic Geometry. Proceedings, 1981. Edited by J.M. Aroca, R. Buchweitz, M. Giusti, and M. Merle. X, 500 pages. 1982.

Vol. 962: Category Theory. Proceedings, 1981. Edited by K.H. Kamps, D. Pumplün, and W. Tholen, XV, 322 pages. 1982.

Vol. 963: R. Nottrot, Optimal Processes on Manifolds. VI, 124 pages. 1982.

Vol. 964: Ordinary and Partial Differential Equations. Proceedings, 1982. Edited by W.N. Everitt and B.D. Sleeman. XVIII, 726 pages. 1982.

Vol. 965: Topics in Numerical Analysis. Proceedings, 1981. Edited by P.R. Turner. IX, 202 pages. 1982.

Vol. 966: Algebraic K-Theory. Proceedings, 1980, Part I. Edited by R.K. Dennis. VIII, 407 pages. 1982.

Vol. 967: Algebraic K-Theory. Proceedings, 1980. Part II. VIII, 409 pages. 1982.

Vol. 968: Numerical Integration of Differential Equations and Large Linear Systems. Proceedings, 1980. Edited by J. Hinze. VI, 412 pages. 1982.

Vol. 969: Combinatorial Theory. Proceedings, 1982. Edited by D. Jungnickel and K. Vedder. V, 326 pages. 1982.

Vol. 970: Twistor Geometry and Non-Linear Systems. Proceedings, 1980. Edited by H.-D. Doebner and T.D. Palev. V, 216 pages. 1982.

Vol. 971: Kleinian Groups and Related Topics. Proceedings, 1981. Edited by D.M. Gallo and R.M. Porter. V, 117 pages. 1983.

Vol. 972: Nonlinear Filtering and Stochastic Control. Proceedings, 1981. Edited by S.K. Mitter and A. Moro. VIII, 297 pages. 1983.

Vol. 973: Matrix Pencils. Proceedings, 1982. Edited by B. Kågström and A. Ruhe. XI, 293 pages. 1983.

Vol. 974: A. Draux, Polynômes Orthogonaux Formels – Applications. VI, 625 pages. 1983.

Vol. 975: Radical Banach Algebras and Automatic Continuity. Proceedings, 1981. Edited by J.M. Bachar, W.G. Bade, P.C. Curtis Jr., H.G. Dales and M.P. Thomas. VIII, 470 pages. 1983.

Vol. 976: X. Fernique, P.W. Millar, D.W. Stroock, M. Weber, Ecole d'Eté de Probabilités de Saint-Flour XI – 1981. Edited by P.L. Hennequin. XI, 465 pages. 1983.

Vol. 977: T. Parthasarathy, On Global Univalence Theorems. VIII, 106 pages. 1983.

Vol. 978: J. Ławrynowicz, J. Krzyż, Quasiconformal Mappings in the Plane. VI, 177 pages. 1983.

Vol. 979: Mathematical Theories of Optimization. Proceedings, 1981. Edited by J.P. Cecconi and T. Zolezzi. V, 268 pages. 1983.

Vol. 980: L. Breen. Fonctions thêta et théorème du cube. XIII, 115 pages. 1983.

Vol. 981: Value Distribution Theory. Proceedings, 1981. Edited by I. Laine and S. Rickman. VIII, 245 pages. 1983.

Vol. 982: Stability Problems for Stochastic Models. Proceedings, 1982. Edited by V. V. Kalashnikov and V. M. Zolotarev. XVII, 295 pages. 1983.

Vol. 983: Nonstandard Analysis-Recent Developments. Edited by A. E. Hurd. V, 213 pages. 1983.

Vol. 984: A. Bove, J. E. Lewis, C. Parenti, Propagation of Singularities for Fuchsian Operators. IV, 161 pages. 1983.

Vol. 985: Asymptotic Analysis II. Edited by F. Verhulst. VI, 497 pages. 1983.

Vol. 986: Séminaire de Probabilités XVII 1981/82. Proceedings. Edited by J. Azéma and M. Yor. V, 512 pages. 1983.

Vol. 987: C. J. Bushnell, A. Fröhlich, Gauss Sums and p-adic Division Algebras. XI, 187 pages. 1983.

Vol. 988: J. Schwermer, Kohomologie arithmetisch definierter Gruppen und Eisensteinreihen. III, 170 pages. 1983.

Vol. 989: A. B. Mingarelli, Volterra-Stieltjes Integral Equations and Generalized Ordinary Differential Expressions. XIV, 318 pages. 1983.

Vol. 990: Probability in Banach Spaces IV. Proceedings, 1982. Edited by A. Beck and K. Jacobs. V, 234 pages. 1983.

Vol. 991: Banach Space Theory and its Applications. Proceedings, 1981. Edited by A. Pietsch, N. Popa and I. Singer. X, 302 pages. 1983.

Vol. 992: Harmonic Analysis, Proceedings, 1982. Edited by G. Mauceri, F. Ricci and G. Weiss. X, 449 pages. 1983.

Vol. 993: R. D. Bourgin, Geometric Aspects of Convex Sets with the Radon-Nikodým Property. XII, 474 pages. 1983.

Vol. 994: J.-L. Journé, Calderón-Zygmund Operators, Pseudo-Differential Operators and the Cauchy Integral of Calderón. VI, 129 pages. 1983.

Vol. 995: Banach Spaces, Harmonic Analysis, and Probability Theory. Proceedings, 1980–1981. Edited by R.C. Blei and S.J. Sidney. V, 173 pages. 1983.

Vol. 996: Invariant Theory. Proceedings, 1982. Edited by F. Gherardelli. V, 159 pages. 1983.

Vol. 997: Algebraic Geometry – Open Problems. Edited by C. Ciliberto, F. Ghione and F. Orecchia. VIII, 411 pages. 1983.

Vol. 998: Recent Developments in the Algebraic, Analytical, and Topological Theory of Semigroups. Proceedings, 1981. Edited by K.H. Hofmann, H. Jürgensen and H. J. Weinert. VI, 486 pages. 1983.

Vol. 999: C. Preston, Iterates of Maps on an Interval. VII, 205 pages. 1983.

Vol. 1000: H. Hopf, Differential Geometry in the Large, VII, 184 pages. 1983.

Vol. 1001: D.A. Hejhal, The Selberg Trace Formula for PSL(2, ℝ). Volume 2. VIII, 806 pages. 1983.

Vol. 1002: A. Edrei, E.B. Saff, R.S. Varga, Zeros of Sections of Power Series. VIII, 115 pages. 1983.

Vol. 1003: J. Schmets, Spaces of Vector-Valued Continuous Functions. VI, 117 pages. 1983.

Vol. 1004: Universal Algebra and Lattice Theory. Proceedings, 1982. Edited by R. S. Freese and O.C. Garcia. VI, 308 pages. 1983.

Vol. 1005: Numerical Methods. Proceedings, 1982. Edited by V. Pereyra and A. Reinoza. V, 296 pages. 1983.

Vol. 1006: Abelian Group Theory. Proceedings, 1982/83. Edited by R. Göbel, L. Lady and A. Mader. XVI, 771 pages. 1983.

Vol. 1007: Geometric Dynamics. Proceedings, 1981. Edited by J. Palis Jr. IX, 827 pages. 1983.

Vol. 1008: Algebraic Geometry. Proceedings, 1981. Edited by J. Dolgachev. V, 138 pages. 1983.

Vol. 1009: T. A. Chapman, Controlled Simple Homotopy Theory and Applications. III, 94 pages. 1983.

Vol. 1010: J.-E. Dies, Chaînes de Markov sur les permutations. IX, 226 pages. 1983.

Vol. 1011: J.M. Sigal. Scattering Theory for Many-Body Quantum Mechanical Systems. IV, 132 pages. 1983.

Vol. 1012: S. Kantorovitz, Spectral Theory of Banach Space Operators. V, 179 pages. 1983.

Vol. 1013: Complex Analysis – Fifth Romanian-Finnish Seminar. Part 1. Proceedings, 1981. Edited by C. Andreian Cazacu, N. Boboc, M. Jurchescu and I. Suciu. XX, 393 pages. 1983.

Vol. 1014: Complex Analysis – Fifth Romanian-Finnish Seminar. Part 2. Proceedings, 1981. Edited by C. Andreian Cazacu, N. Boboc, M. Jurchescu and I. Suciu. XX, 334 pages. 1983.

Vol. 1015: Equations différentielles et systèmes de Pfaff dans le champ complexe – II. Seminar. Edited by R. Gérard et J.P. Ramis. V, 411 pages. 1983.

Vol. 1016: Algebraic Geometry. Proceedings, 1982. Edited by M. Raynaud and T. Shioda. VIII, 528 pages. 1983.

Vol. 1017: Equadiff 82. Proceedings, 1982. Edited by H. W. Knobloch and K. Schmitt. XXIII, 666 pages. 1983.

Vol. 1018: Graph Theory, Łagów 1981. Proceedings, 1981. Edited by M. Borowiecki, J.W. Kennedy and M.M. Sysło. X, 289 pages. 1983.

Vol. 1019: Cabal Seminar 79–81. Proceedings, 1979–81. Edited by A. S. Kechris, D. A. Martin and Y. N. Moschovakis. V, 284 pages. 1983.

Vol. 1020: Non Commutative Harmonic Analysis and Lie Groups. Proceedings, 1982. Edited by J. Carmona and M. Vergne. V, 187 pages. 1983.

Vol. 1021: Probability Theory and Mathematical Statistics. Proceedings, 1982. Edited by K. Itô and J.V. Prokhorov. VIII, 747 pages. 1983.

Vol. 1022: G. Gentili, S. Salamon and J.-P. Vigué. Geometry Seminar "Luigi Bianchi", 1982. Edited by E. Vesentini. VI, 177 pages. 1983.

Vol. 1023: S. McAdam, Asymptotic Prime Divisors. IX, 118 pages. 1983.

Vol. 1024: Lie Group Representations I. Proceedings, 1982–1983. Edited by R. Herb, R. Lipsman and J. Rosenberg. IX, 369 pages. 1983.

Vol. 1025: D. Tanré, Homotopie Rationnelle: Modèles de Chen, Quillen, Sullivan. X, 211 pages. 1983.

Vol. 1026: W. Plesken, Group Rings of Finite Groups Over p-adic Integers. V, 151 pages. 1983.

Vol. 1027: M. Hasumi, Hardy Classes on Infinitely Connected Riemann Surfaces. XII, 280 pages. 1983.

Vol. 1028: Séminaire d'Analyse P. Lelong – P. Dolbeault – H. Skoda. Années 1981/1983. Edité par P. Lelong, P. Dolbeault et H. Skoda. VIII, 328 pages. 1983.

Vol. 1029: Séminaire d'Algèbre Paul Dubreil et Marie-Paule Malliavin. Proceedings, 1982. Edité par M.-P. Malliavin. V, 339 pages. 1983.

Vol. 1030: U. Christian, Selberg's Zeta-, L-, and Eisensteinseries. XII, 196 pages. 1983.

Vol. 1031: Dynamics and Processes. Proceedings, 1981. Edited by Ph. Blanchard and L. Streit. IX, 213 pages. 1983.

Vol. 1032: Ordinary Differential Equations and Operators. Proceedings, 1982. Edited by W. N. Everitt and R. T. Lewis. XV, 521 pages. 1983.

Vol. 1033: Measure Theory and its Applications. Proceedings, 1982. Edited by J. M. Belley, J. Dubois and P. Morales. XV, 317 pages. 1983.

Vol. 1034: J. Musielak, Orlicz Spaces and Modular Spaces. V, 222 pages. 1983.